Temporal and Compressive Properties of the Normal and Impaired Auditory System

Ralph Peter Derleth

Bibliotheks- und Informationssystem der Universität Oldenburg
1999

Verlag/Druck/	Bibliotheks- und Informationssystem
Vertrieb:	der Carl von Ossietzky Universität Oldenburg
	(BIS) - Verlag -
	Postfach 25 41, 26015 Oldenburg
	Tel.: 0441/798 2261, Telefax: 0441/798 4040
	e-mail: verlag@bis.uni-oldenburg.de

ISBN 3-8142-0695-9

Contents

Preface 7

1   General introduction 9

2   **Amplitude modulation detection in normal and hearing-impaired listeners** 15

2.1   Introduction 17
2.2   Experiment I: Modulation detection with stochastic carriers 20
2.2.1   Method 20
2.2.1.1   Procedure and subjects 20
2.2.1.2   Apparatus and stimuli 21
2.2.1.3   Simulations 23
2.2.2   Results 23
2.2.2.1   Modulation detection with narrowband carriers 23
2.2.2.2   Level dependence of modulation detection 25
2.2.3   Simulations 26
2.3   **Experiment II: Intensity discrimination and modulation detection with sinusoidal carriers** 28
2.3.1   Method 28
2.3.1.1   Procedure and subjects 28
2.3.1.2   Apparatus and stimuli 29
2.3.2   Results 30
2.3.2.1   Increment- and modulation-detection thresholds 30
2.3.3   Simulations 33
2.4   **Discussion** 35
2.4.1   Stochastic carrier 35
2.4.2   Deterministic conditions 36
2.5   **Conclusions** 38

3 Modeling loudness matching and modulation matching 39

- 3.1 **Introduction** 40
- 3.2 **Model** 42
- 3.2.1 Adaptation stage for normal hearing 43
- 3.2.2 Adaptation stage for sensorineural impaired hearing 44
- 3.3 **Loudness matching for sinusoidal stimuli** 46
- 3.3.1 Method 46
- 3.3.1.1 Experimental procedure and stimuli 46
- 3.3.1.2 Simulations 46
- 3.3.2 Results 47
- 3.4 **Modulation matching** 49
- 3.4.1 Method 49
- 3.4.1.1 Experimental procedure and stimuli 49
- 3.4.1.2 Simulations 49
- 3.4.2 Results 50
- 3.4.3 Analysis of model predictions 52
- 3.5 **Summary and discussion** 54
- 3.6 **Conclusions** 56

4 Modeling temporal and compressive properties of the normal and impaired auditory system 57

- 4.1 **Introduction** 59
- 4.2 **Models** 62
- 4.2.1 Model 1 62
- 4.2.2 Model 2 65
- 4.2.3 Model 3 66
- 4.2.4 Simulations 69
- 4.3 **Results** 70
- 4.3.1 Loudness functions 70
- 4.3.2 Frequency selectivity as a function of level 71
- 4.3.3 Forward masking 73
- 4.3.4 Modulation detection 77
- 4.3.5 Modulation matching 79
- 4.3.5.1 Analysis of model predictions 81

4.4	Discussion	83
4.5	Conclusions	87

5 On the role of envelope modulation processing for spectral masking effects ... 89

5.1	Introduction	90
5.2	Experiment I: Masking patterns for sinusoidal and noise maskers and sinusoidal and noise signals	93
5.2.1	Method	93
5.2.1.1	Procedure and Stimuli	93
5.2.1.2	Simulations	94
5.2.2	Results	95
5.2.2.1	Masking patterns obtained with a tone masker	95
5.2.2.2	Masking patterns obtained with a noise masker	98
5.3	Experiment II: Notched-noise masking	101
5.3.1	Method	101
5.3.1.1	Procedure and Subjects	101
5.3.1.2	Apparatus and Stimuli	101
5.3.1.3	Simulations	102
5.3.2	Results	102
5.4	Discussion	105
5.5	Conclusions	107

6 Summary and concluding remarks ... 109

A Auditory processing model ... 115

B Inter- and intraindividual differences ... 119

C Effects of template level ... 121

References ... 123

Danksagung ... 131

Preface

The distorted processing of sounds in the ears of hearing-impaired patients is one of the most complex and yet not at all understood principal problems in hearing research with many potential applications: Although nowadays "intelligent" digital hearing aids have more and more processing power within an incredibly small volume, their software and fitting procedures are still limited by our very little knowledge on how the impaired ear works and how it can be supported by signal processing in an optimum way.

To solve this problem, Ralph Peter Derleth uses in his thesis the classical approach of modern physics: Interaction between experiment and theory to create a better understanding of the underlying system. The experiments conducted by him are concerned with the perception of amplitude modulations as a function of stimulus level, modulation rate, and carrier bandwidth both for normal and hearing-impaired listeners. The models employed by him are modifications of the "Oldenburg model" for the effective signal processing in the auditory system that have been treated in a series of dissertations by Dau (1996) and Verhey (1999) from the same publisher. This approach may be termed as "computational psychophysics": The computer simulates the actual performance of the human observer by assuming an imperfect transformation of the acoustical stimulus into the "internal auditory representation" which in itself is assumed to be processed in an optimum way. Although this approach is remarkably simple and uses comparatively few assumptions about the auditory system, it has successfully been applied to predict a variety of psychoacoustical effects in normal listeners as well as some applications in speech processing. The prominent achievement of the current thesis by Ralph Peter Derleth is now to extend this work to hearing-impaired listeners and to give fascinating new insights and interpretations on how the impaired ear works.

Within a series of three independent chapters (each of which has been prepared as a publication in an international acoustical journal), Ralph Peter Derleth can show that the "effective" compression apparent in the normal auditory system is divided up into a fast, peripheral compressive process (presumably in the cochlea) and a slower, more central adaptation process which is responsible for temporal integration processes and effects such as forward and backward masking. Since too much compression is not a good thing, some intermediate expansion stage has to be included which linearizes the peripheral auditory system in normal-hearing listeners and produces the famous "recruitment" phenomenon in hearing-impaired listeners where the fast-acting peripheral compressive property is lost. This new interpretation of sensorineural hearing impairment as a loss of the "effective" linearity of the peripheral auditory system derived in this thesis opens the scene for a variety of new theories on normal and impaired hearing as well as future generations of "intelligent" hearing aids. But yet another important aspect is given in the current thesis (Chapter 5): Based on his experiments and models on modulation perception, Ralph Peter Derleth solves the riddle of the "asymmetry of masking" paradigm, i.e., the problem that a sinusoid exerts less masking on a narrow-band noise than vice versa, although both signals have approximately the same spectral masking pattern. I don't want to tell you the solution of this riddle right away. Please read yourself! Just one hint: The temporal structures of both types of signals are rather different. Possible consequences of Ralph Peter Derleth's accurate modeling might be more efficient speech and audio coding strategies that may utilize these effects for low-bit rate, high-quality acoustical transmission.

Although Ralph Peter Derleth has been the 16th Ph.D. candidate from our group and continues the sequence of excellent psychoacousticians graduating from the interdisciplinary graduate school Psychoacoustics, he – and, of course, his work – is very special: Not only his pioneering work in modeling the impaired auditory system and his patience to deal with patients qualifies him or the fact, that he is the first "real Oldenburger" from our Göttingen-based group that moved to Oldenburg six years ago. Among other virtues, his endurance in dealing with such difficult matters as apparently never-working models or non-cooperative computers and other hardware, his talent in organizing summer schools and his always supportive attitude and open-mindedness really makes him a special person who is just starting a great career. Just read yourself and you will be convinced, too!

Oldenburg, November 1999

Birger Kollmeier

Chapter 1
General introduction

The auditory system is a complex structure which performs a transformation of sound energy into an internal auditory percept which provides us a variety of useful informations further exploited by cognitive processes. The internal percept includes environmental information and allows us among other things to communicate with other human beings. As long as this transformation works in its normal manner, it is hardly recognized in a conscious manner. If any kind of hearing impairment occurs, however, this transformation between sound energy and percept is noticeably altered. In dependence on the processing structure where the impairment takes place, a conductive hearing impairment (outer and middle ear) is differentiated from a sensorineural hearing impairment (cochlear and higher stages in the auditory pathway). The effect of a conductive hearing impairment is simply a frequency-specific reduction in sound energy that enters the cochlea. Therefore, a normal perception of sound can be achieved by a frequency-specific increase in sound energy for subjects suffering from this kind of hearing impairment. For sensorineural hearing impairment, however, the situation is by far more complex. Usually, this kind of impairment is primary due to a lesion within the cochlea although sensorineural hearing loss can involve structures other than the cochlea. Because of the complex interaction of the inner ear structures, any dysfunction of one part of the system affects the processing of sound in a large variety of manners. The variability across subjects in individual hearing impairment is therefore very high and it is still unclear if and how a normal perception of sound can be achieved for these subjects by means of „intelligent" signal-processing hearing aids.

In principle, there are two approaches for modeling the perception abilities of the human auditory system. The first one represents solely the „effective" signal processing of the auditory system as it can be observed in psychoacoustical experiments. Within such a model approach, some processing stages may be associated with physiological observed structures, others may not. The main features of such a model approach are a relatively simple model structure, few parameters and only low computational needs. The processing model (Dau et al., 1996a+b, 1997a+b) for simulating normal hearing belongs to this class of models. The other class of models tries to simulate the physiologically observed realizations of some aspects of the auditory system as close as possible. This may lead to a large number of parameters and a very complicated model structure. The computational needs are very high, often too high for the complete simulation of psychoacoustical measurements (because often stimuli with relatively long duration's (hundreds of milliseconds) and measurement procedures which require a large number of stimulus presentations (adaptive n-AFC procedure) are used). The goal is to incorporate those physiologically observed results in an „effective" way within the processing model which most likely have a strong influence on signal processing in the auditory system.

This thesis is concerned with modeling the „effective" signal processing in the auditory system for hearing-impaired listeners. This subject is highly desirable for various reasons: First, hearing impairment can be considered as a variation of the functioning of the auditory system which carries information about the construction of the auditory system and thus helps to understand the normal auditory signal processing. Second, such a model would help to establish the relations between different audiological parameters that are derived from audiological tests with hearing-impaired patients in clinics or research. Based on this knowledge, more refined signal processing methods can be derived for „intelligent" hearing aids that aim at compensating for the reduced auditory capabilities. Moreover, such a model would be helpful to assess hearing aid strategies in an „objective" way that does not require exhaustive measurements with individual hearing-impaired listeners.

Most models of the impaired auditory system concentrate on certain aspects of auditory functions such as, e.g. reduced frequency selectivity (Florentine et al., 1980), reduced temporal resolution (Florentine and Buus, 1984), distorted intensity or loudness perception (Moore, 1995, Moore 1997, Launer et al., 1997) and binaural interaction (Colburn et al., 1987).

Considerable attention has been given to the question of which aspects of hearing impairment can be considered as „primary factors" such as, e.g., the loss of sensitivity caused by a reduced function of inner hair cells and the loss of dynamic compression caused by a deficit in outer hair cell functioning (cf. Moore, 1995, Borg et al., 1995). They differ from „secondary factors" (such as, e.g., temporal resolution, frequency resolution and binaural interaction) that might be a direct consequence of the alterations in the primary factors. Florentine et al. (1980) and Moore (1995), for example, have argued that a broadening of the peripheral auditory filters is a direct consequence of the hearing loss that has to be accounted for when modeling loudness summation across frequency. On the contrary, Launer et al. (1997) and Launer (1995) have argued that loudness summation across frequency can be adequately modeled by assuming the same sound pressure level dependence of auditory filter-bands as in normal-hearing listeners but assuming an abnormal loss in the compressive nonlinearity of the auditory system (caused by the dysfunction of the outer hair cells). These relative significance of different factors in auditory signal processing for hearing-impaired listeners is important in order to determine what kind of audiological parameters would be necessary and sufficient to characterize a sensorineural hearing impairment.

Another class of models considers the „effective" processing in the auditory system and tries to predict as many psychoacoustical properties as possible with a minimum set of assumptions and parameters. Such a model was presented by Dau et al. (1996a+b, 1997a+b). It consists of several preprocessing stages and an optimal detector as the decision device. One of the main advantages of this model is that it can simulate the psychoacoustical measurement procedures and therefore predict experimental results not only qualitatively but also quantitatively. It was shown by the authors that this model accounts for many experimental results concerning the temporal- and the spectral detection abilities of the healthy auditory system, e.g. intensity discrimination, simultaneous and forward masking, modulation detection and modulation masking, test-tone integration. Other authors further presented model predictions on notched-noise data (Verhey, 1998) and some effects observed with experiments on comodulation masking release (Verhey et al., 1999). The model has been further developed to predict different properties such as, e.g., speech perception (Holube and Kollmeier, 1996), speech transmission quality (Hansen and Kollmeier, 1997). Modeling efforts have also been undertaken to present a model which accounts for binaural effects (Zerbs, 1999). Moreover, the preprocessing stages of the model are used in speech recognition systems (Tchorz and Kollmeier, 1999).

In an attempt to also predict the effect of hearing impairment, Holube and Kollmeier (1996) introduced an elevated threshold, broadened auditory filters and an increase in auditory time constants. Although the model could predict psychoacoustical data (forward-masking and notched-noise data) in hearing impaired listeners, the modification of filter-bands and time constants were not salient in predicting the deteriorated performance in speech intelligibility in quiet and noise. In addition, their approach was not able to describe the „recruitment-phenomenon" in hearing-impaired listeners, i.e., the distorted relation between loudness perception and stimulus intensity. That is, the absolute threshold of the stimulus is elevated in comparison to a normal-hearing subject but the stimulus level which is judged as uncomfortable loud remains nearly unchanged to that of a normal-hearing listener. Moreover, the assumed mechanism of a reduced temporal resolution for hearing impaired subjects was contradicted by a study on temporal resolution using the method of measuring the temporal modulation transfer function (TMTF) (Hohmann, 1993).

Within this thesis it is assumed that the recruitment effect is one of the primary factors influencing the altered perception in sensorineural hearing-impaired subjects. The physiological correlate which is assumed to be a main factor for the recruitment effect is the loss of the fast-acting level-dependent compression observed in the normal functioning mammalian cochlea (Recio et al., 1998). Depending on the degree if impairment, the cochlea reacts less compressive or even linear (Ruggero, 1992). Several modeling approaches which realize a difference in compressive properties of the model version for simulating normal and impaired hearing are presented. The modeling strategy is oriented on the following principles: (i) the modeling approach is based on an earlier model which has proven to be able to account for a wide variety of psychoacoustical data obtained with normal hearing subjects, (ii) taking into account what we know from physiology, (iii) the use of few parameters for describing a sensorineural hearing impairment.

Chapter 2 of this thesis deals with the temporal processing properties of the healthy and impaired auditory system. Experiments on modulation masking, modulation detection and intensity discrimination obtained with normal-hearing and hearing-impaired listeners are described which investigate if and how the temporal properties of the model have to be altered in order to describe the impaired auditory system. Often, broadband noise has been used as the carrier in modulation detection and modulation masking experiments (e.g. Viemeister, 1979). The use of broadband-noise carrier, however, precludes investigation of temporal processing in the different frequency regions which is especially interesting in case of a frequency-specific hearing loss. Therefore, either narrowband-noise or sinusoids were used as the carrier in the present study. The spectral content of each stimulus was restricted to one critical band of normal-hearing listeners to preclude spectral information as a decision cue. The level of the carrier was chosen to produce an equal-loud sensation in both groups of subjects with normal hearing and impaired hearing to rule out the influence of altered loudness perception and to make sure that all parts of the stimulus were audible to the listeners.

Chapter 3 deals with processing properties of the healthy and impaired auditory system on perceived sensations, like loudness and the perceived depth of modulation. The model was used so far for simulating detection experiments and thus, a correlate of the above sensations within the model had to be postulated. In case of the loudness of a stimulus, the mean excitation in the model produced by the stimulus (averaged over a certain time) is considered. In case of the perceived modulation depth, the modulation-energy (at a certain processing stage) within the model produced by the stimulus is considered. Experiments on loudness matching and modulation matching obtained with unilateral sensorineural hearing-impaired listeners (Moore et al., 1996) are simulated with two model approaches which realize less compression in the model version for impaired hearing than in the model version for normal hearing.

Chapter 4 concentrates on the question if and how a level-dependent fast-acting compression -similar to the one observed in the healthy cochlea- can be realized within the model. Three model approaches which incorporate a level-dependent difference in compressive properties between the model version for normal hearing and the model version for impaired hearing are presented. The processing properties of the models are compared under the aspect of predicted differences between the model version for normal hearing and the model version for impaired hearing. Experiments on perceived loudness and forward-masking are presented. In addition, the modulation transfer characteristics of the different model approaches are compared which allow to clarify to which extent the models account for modulation matching data.

In the previous chapters, the importance of processing temporal cues (e.g. amplitude modulations) for modeling the performance of normal-hearing and hearing-impaired listeners in detection-experiments and experiments involving perceived sensations became apparent. Chapter 5 considers if these cues can be exploited to solve an old paradoxon from psychoacoustic: the asymmetry of masking. Simulations of experiments on masking patterns (Moore et al., 1998) obtained with sinusoidal and narrowband signals and sinusoidal and narrowband maskers are presented. In addition to the simulations obtained with the model for normal hearing which processes also higher-frequency modulations, simulations obtained with a model which basically works as an energy detector are presented. Own experimental data and simulations obtained with the notched-noise paradigm using the bandwidth of the masking noise-bands as parameter are also presented.

Taken together, this thesis is an attempt to tackle some of the most important and still unresolved issues in auditory perception: temporal acuity, modulation perception and hearing impairment.

Chapter 2
Amplitude modulation detection in normal and hearing-impaired listeners[1]

Abstract

Modulation-detection and intensity-discrimination experiments are measured for five normal-hearing and four sensorineural hearing-impaired listeners and simulated with the modulation-filterbank model proposed by Dau et al. (1997a). The temporal modulation transfer function (TMTF) is measured for narrow-band noise carriers ($\Delta f = 16$ or 200 Hz) centered at 2 kHz. In addition, the intensity-discrimination threshold and the modulation-detection threshold for a modulation frequency of $f_m = 16$ Hz is measured for a sinusoidal carrier for three of the five normal-hearing listeners and a different group of four sensorineural hearing-impaired subjects. The carrier level is individually chosen to produce the same perceived loudness impression for each listener. The experiments on modulation detection both with stochastic and deterministic carriers demonstrate that the processing of amplitude modulations is the same for normal-hearing and sensorineural hearing-impaired listeners. The mean threshold difference between experimental results on intensity discrimination and modulation detection with a sinusoidal carrier is the same for both groups of listeners. These findings indicate that a loss of peripheral (cochlea) compression does not affect the

---

[1] Modified version of the paper „Amplitude modulation detection in normal and hearing-impaired listeners", written together with Torsten Dau and Birger Kollmeier, submitted to J. Acoust. Soc. Am.

ability of the auditory system to detect modulations. Instead, a general loss of sensitivity in both experimental conditions can be observed for some hearing-impaired subjects. This general loss appears to be not directly related to the amount of sensorineural hearing impairment. The proposed processing model is capable of quantitatively modeling most aspects of the modulation-masking, modulation-detection and intensity-discrimination experiments described. The large difference in thresholds for intensity discrimination and modulation detection, respectively, is accounted for by a retrocochlear adaptive compression, realized in the adaptation stage of the model. Stationary stimuli are compressed whereas modulations are transformed less compressive.

2.1 Introduction

There is a large variety of hearing impairments which result in an even wider variety of altered perception and detection abilities of the impaired auditory system. A major effect found with sensorineural hearing impairment is the recruitment phenomenon. It is characterized by an elevated absolute threshold and near-to-normal uncomfortable loudness level (UCL), i.e., only a reduced dynamic range can be used for the perception of signals. It is assumed (e.g. Moore, 1995) and widely accepted that the recruitment phenomenon is based on a loss of compressive nonlinearity present in the healthy cochlea (Ruggero and Rich, 1991). The comparison of psychoacoustical perception abilities between normal-hearing and sensorineural hearing-impaired listeners is largely influenced by this reduced dynamic range. In the literature, psychoacoustical abilities are often compared at the same sound pressure level (SPL), or the same sensation level (SL) above the individual absolute threshold. An alternative approach that takes the reduced dynamic range into account, is the comparison at levels producing the same loudness perception for average normal-hearing and the individual hearing-impaired listeners. This approach is adopted in the current study since it attempts to compare the performance of normal-hearing and hearing-impaired listeners at a similar output activation of the peripheral auditory system.

    Temporal resolution in the auditory system, or the ability to resolve dynamic acoustic cues, plays an important role for the processing of complex sounds such as speech. A hearing impairment often affects the listeners ability to understand speech, especially in situations when environmental noise is apparent or several speakers communicate simultaneously. Several researchers have examined the question if the reduced performance in speech recognition is caused by a reduced temporal resolution of sensorineural hearing-impaired listeners (Dreschler and Plomp, 1980, 1985, Festen and Plomp, 1983, Tyler et al., 1982, Glasberg and Moore, 1989, van Rooij and Plomp, 1990, Lutman, 1991). However, impaired temporal resolution usually occurs together with other impaired functions making it difficult to isolate the specific effect of this factor on speech recognition. Some studies have reported significant correlation between temporal resolution and speech intelligibility (e.g. Tyler et al., 1982) while other have found no significant correlation (e.g. Festen and Plomp, 1983). Experiments with temporally smeared speech envelopes (Drullman et al., 1993, Hou and Pavlovic, 1994) suggest that the ability to follow slow envelope fluctuations has significant influence on speech perception for normal hearing subjects.

One general psychoacoustical measure of temporal resolution is the temporal modulation transfer function (TMTF), i.e., the sensitivity for modulations as a function of the modulation frequency. In studies with normal-hearing listeners, wide-band noise is often used as a carrier stimulus in order to prevent subjects from using changes in the overall spectrum as a detection cue (Viemeister, 1979). However, the hearing loss in hearing-impaired subjects varies considerably with frequency so that their temporal resolution has to be measured separately for the frequency regions with different absolute thresholds. Hence, narrowband stimuli should be used as the carrier to obtain a frequency-specific comparison of temporal resolution of hearing-impaired subjects with that of normal-hearing listeners. Conversely, it is questionable if the use of wideband noise carriers in connection with a frequency-specific hearing loss (Bacon and Viemeister, 1985) is able to detect any changes in temporal resolution in the impaired frequency region. Moore et al. (1992) showed for an octave-wide noise carrier centered at 2 kHz that the temporal resolution for hearing-impaired listeners is as well as that for normal-hearing listeners. In order to restrict the experiment to an even narrower spectral region, all spectral components of the stimuli lie within one critical band in the current study.

In the first experiment, the detection thresholds for amplitude modulation of a sound is measured as a function of the modulation rate for narrowband noise as the carrier. In this experiment, the inherent statistics of the carrier is the limiting factor for the detection of the test modulation. The data of normal-hearing subjects show a certain frequency selectivity for modulations (Dau et al., 1997a). Measurements with normal-hearing and sensorineural hearing-impaired listeners should therefore reveal if this modulation-frequency selectivity is altered by the loss of compression in the cochlear. The modulation-frequency selectivity has proven to be an important aspect for modeling the performance of normal-hearing listeners in psychoacoustical tasks (Dau et al., 1997a+b, Verhey, 1998). The knowledge of an altered modulation-frequency selectivity is therefore expected to be critical for modeling the performance of hearing-impaired listeners.

To partial out the influence of an assumed general loss of sensitivity in hearing-impaired subjects on altered temporal resolution, in Exp. II deterministic stimuli (tones) were used as the carrier so that no external statistics

2.1 Introduction

is apparent and the internal noise is the limiting factor for the detection of the test modulation. In addition, the intensity-discrimination threshold is measured to reveal the influence of the internal noise both on intensity discrimination and modulation detection.

2.2 Experiment I: Modulation detection with stochastic carriers

2.2.1 Method

2.2.1.1 Procedure and subjects

Masked thresholds were measured using an adaptive three-interval forced-choice (3IFC) procedure. The masker was presented in three consecutive intervals, separated by silent intervals of 500 ms. In one randomly chosen interval the test stimulus was added to the masker. The subject's task was to specify the interval containing the test stimulus. The threshold was adjusted by a 1-up 2-down algorithm (Levitt,1971), which converges at a stimulus level corresponding to a probability of being correct of 70.7%. The step size was 4 dB at the start of a run and was divided by 2 after every two reversals of the signal level until the step size reached a minimum of 1 dB that was kept fixed during the remainder of the track. Once this step size was achieved, eight reversals were obtained and the median-value of the stimulus levels at these eight reversals was used as the threshold estimate. The subjects received no feedback during the measurements. The procedure was repeated four times for each signal configuration and subject.

Two male and three female normal-hearing listeners (nl) (audiometric thresholds within 10 dB HL, no history of hearing problems) volunteered to participate in the experiment. They were between 24 and 32 years old. Three of them were trained in psychoacoustic measurements. Two male and two female sensorineural hearing-impaired listeners (il) participated at the experiment. All had experience in psychoacoustic measurements. They were aged between 59 and 73 years. Figure 2.1 shows the audiometric thresholds of the hearing-impaired listeners. In the range from 1.5 kHz to 3 kHz all hearing-impaired subjects show a flat hearing loss (il1, il3, il4) or a mild downward-sloping hearing loss (il2) of 50-80 dB HL. The threshold difference between bone- and air- conduction was smaller than 10 dB in the specified range. In case of an asymmetric hearing loss always the better ear was tested. The hearing-impaired listeners were paid for their participation on an hourly basis.

There are several experimental methods to reveal the connection between the physical level of a sound and its perceived loudness. The basic

2.2 Experiment I: Modulation detection with stochastic carriers

idea is to sort signals in verbal categories representing the loudness percept (Pascoe, 1978; Hellbrück and Moser, 1985; Allen et al., 1990). In the present study, a loudness scaling procedure adapted by Launer (1995) was used to reveal the levels of perceptive equality. The loudness was measured in categorical units (CU) using a one-step categorical scaling procedure with eleven categories ranging from „unhörbar" (CU 0, „inaudible") to „zu laut" (CU 50, „too loud") (see Tab. 2.1 for the remaining categories). Each signal was presented two times in random order (with restricted stepsize) at eight stimulus levels. The levels were equispaced on a dB scale covering the entire dynamic range defined by audiometric threshold and uncomfortable loudness level (UCL). For details of the procedure see Launer (1995) and Brand et al. (1999).

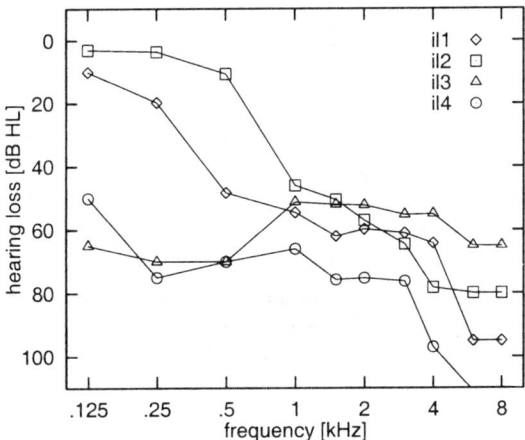

Fig. 2.1: Audiometric thresholds for four sensorineural hearing-impaired listeners. The ordinate indicates the hearing loss in dB HL and the abscissa represents the frequency of the sinusoidal test signal.

2.2.1.2 Apparatus and stimuli

All acoustic stimuli were digitally generated at a sampling frequency of 12 kHz. The stimuli were transformed to analog with a two-channel 16-bit D/A converter (ARIEL DSP32C), attenuated and low-pass filtered at 6 kHz with a programmable amplifier and presented monaurally via headphones (Sennheiser HDA200) in a double-walled soundproof booth (IAC-1600). Signal generation and presentation of trials were controlled by a computer using the signal-processing software package SI (developed at the University Göttingen). The subjects responded via a PC keyboard or, in case of the categorical loudness scaling procedure, via a handhold touch screen (Epson ETH-10) which displayed the categories after each presentation of the stimulus.

The carrier used in this experiment was a bandpass-filtered Gaussian noise of 5 s duration. The bandwidth of the noise was either 200 or 16 Hz, centered symmetrically on a linear frequency scale around 2 kHz. Each noise interval had a duration of 1 s and was randomly cut from the masker noise to perform a running-noise experiment. The test signal (s(t)) was a sinusoidally amplitude modulated (SAM) noise (see Form. 2.1) where the modulation had the same duration (1 s) as the carrier (c(t)). The modulation frequency (f_{mod}) was 4, 8, 16, 32, 64 and 80 Hz, respectively. For comparison, thresholds for a SAM-sinusoid at 2 kHz of 1 s duration were obtained for modulation frequencies $\geq$ 16 Hz.

$$s(t) = c(t) \cdot (1 + m \cdot \cos(2\pi \cdot f_{mod} \cdot t)) \qquad (2.1)$$

The bandwidth of the modulated test signal was restricted to 200 Hz after modulation. This was achieved by setting all frequency bins of the FFT-transformed 1s-signal to zero in the frequency domain. Rising and falling $\cos^2$-shaped ramps of 125 ms duration were applied to the stimulus. Finally, each stimulus was adjusted to have the same rms-value. The presentation level was 60 dB SPL for the normal-hearing listeners and was chosen individually for the hearing-impaired listeners such that the perceived loudness equaled the category „mittellaut" (25 categorical units), i.e., loudness category „intermediate". Table 2.1 shows the individual presentation levels in dB SPL corresponding to a given loudness category (same for the 200 Hz-wide and 16 Hz-wide carrier).

categories [CU]	listener nl mean	il1	il2	il3	il4
45 „sehr laut" („very loud")	95	105	105	115	120
35 „laut" („loud")	80	98	93	104	110
25 „mittellaut" („intermediate")	60	88	80	91	99
15 „leise" ("soft")	40	80	65	78	90
5 „sehr leise" („very soft")	15	72	60	63	78

Tab. 2.1: Levels for the 2 kHz-narrowband noise (200 Hz) in dB SPL for the normal-hearing and the hearing-impaired listeners. Different categorical units (CU) correspond to different loudness categories, given in quotes.

2.2.1.3 Simulations

Simulations were performed using the auditory processing model for normal-hearing listeners (Dau et al., 1997a) (for a brief description of the model see appendix A). No specific alterations were introduced within the model for simulating impaired hearing. For predicting thresholds, the same procedure and stimuli as in the experiment were used.

2.2.2 Results

2.2.2.1 Modulation detection with narrowband carriers

Temporal modulation transfer functions (TMTFs) were measured for narrowband noise carriers of the bandwidths 16 and 200 Hz, respectively, centered at 2 kHz in each condition. Fig. 2.2 shows the experimental results for the 200-Hz bandwidth. The ordinate denotes modulation depth at threshold (expressed as 20log(m) in dB), and the abscissa represents modulation frequency. The left panel of Fig. 2.2 shows the thresholds for five normal-hearing subjects. The right panel represents the data for the four sensorineural hearing-impaired subjects. The slope of the threshold patterns as well as the absolute threshold values are very similar for both groups of subjects.

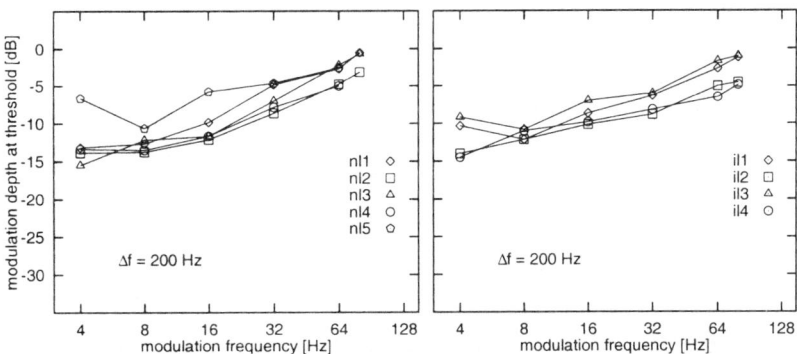

Fig. 2.2: Modulation-detection thresholds of sinusoidal amplitude modulation as a function of the modulation frequency. Data of five normal-hearing subjects (left panel) and of four sensorineural hearing-impaired subjects (right panel) are shown as mean-values (standard deviation < 3.5 dB). The carrier was a 200 Hz-wide running Gaussian noise at a center frequency of 2 kHz. Carrier and modulation duration: 1 s. Level: corresponding to 25 CU (60 dB SPL for normal-hearing listeners, individual levels for hearing-impaired listeners according to Tab. 2.1).

For modulation frequencies above 8 Hz, thresholds increase by about 3 dB per doubling of the modulation frequency. This threshold pattern agrees well with corresponding experimental data by Eddins (1993) who measured the TMTF for normal-hearing subjects for a set of carrier bandwidths. The lowpass-shaped form of the TMTF is similar to the pattern found in „classical" measurements of the TMTF using a broadband noise as the carrier, but exhibits a lower corresponding „cut-off" frequency of approximately 16 Hz compared to 64 Hz found with broadband noise as the carrier (Viemeister, 1979).

Figure 2.3 shows the experimental results for a carrier bandwidth of 16 Hz (solid lines). In addition, thresholds obtained with a 2 kHz sinusoidal carrier are shown (dashed lines). The left panel of Fig. 2.3 shows data for five normal-hearing subjects and the right panel shows the data of four sensorineural hearing-impaired subjects. As in the previous experimental condition, the shape of the threshold pattern as well as the absolute threshold-values are very similar for both groups. For modulation frequencies larger than half the bandwidth of the noise ($f_{mod} > \Delta f/2$), thresholds start to decrease, so that the threshold function exhibits a high-pass characteristic.

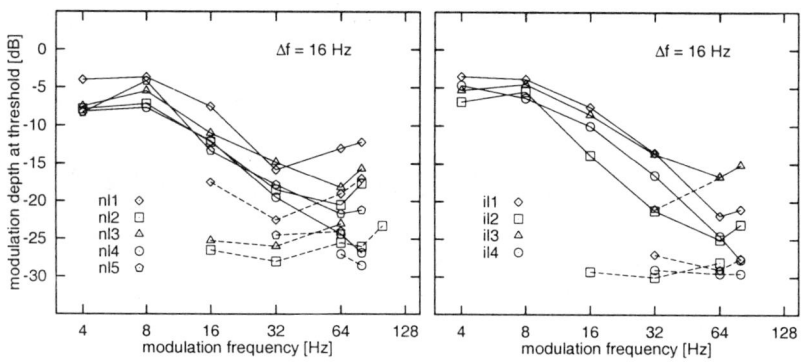

Fig. 2.3: Modulation-detection thresholds of sinusoidal amplitude modulation as a function of the modulation frequency. Data of five normal-hearing subjects (left panel) and four sensorineural hearing-impaired listeners (right panel) are shown as mean-values of four measurements (standard deviation typically < 3.5 dB, see appendix B for details). The carrier was a 16 Hz-wide running Gaussian noise at a center frequency of 2 kHz. Carrier and modulation duration: 1 s. Level: corresponding to 25 CU (60 dB SPL for normal-hearing subjects).

Even at the high modulation rates 64 and 80 Hz, thresholds have not yet approached with those for the sinusoidal carrier, but are at least 2 to 5 dB

2.2 Experiment I: Modulation detection with stochastic carriers

higher. The only exception can be observed for the impaired listener il3, who exhibits similar thresholds for the sinusoidal and the noise carrier at 64 Hz modulation rate.

In the $\Delta f = 200$-Hz condition (Fig. 2.2), the intraindividual and interindividual data both for the normal-hearing and for the hearing-impaired listeners lie within the range of 3.5 dB and do not depend on modulation frequency. This also holds for the interindividual data in the $\Delta f = 16$-Hz condition (Fig. 2.3). However, the interindividual standard deviation increases markedly at high modulation rates for both groups of subjects (see appendix B for details).

2.2.2.2 Level dependence of modulation detection

Figure 2.4 shows the modulation detection thresholds for 4 Hz as a function of the masker level. The ordinate indicates the modulation depth at threshold in dB, and the abscissa represents the perceived loudness in categorical units. The left panel of Fig. 2.4 shows mean data and standard deviations for the $\Delta f = 200$-Hz condition and the right panel shows data for the $\Delta f = 16$-Hz condition. The diamonds indicate data for the five normal-hearing listeners whereas the boxes represent the corresponding data for the

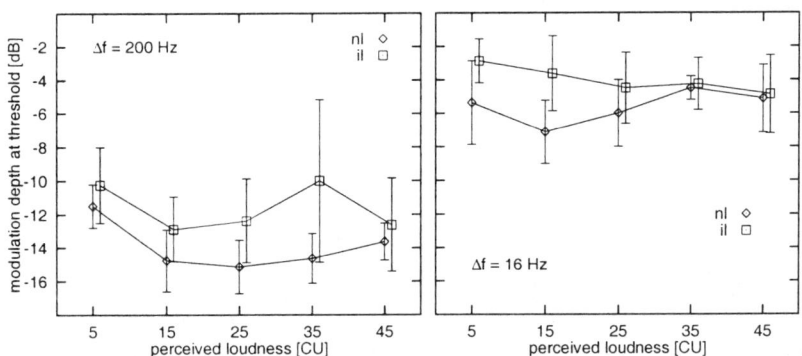

Fig. 2.4: Modulation-detection thresholds for 4 Hz as a function of the perceived loudness. Mean data and standard deviation of five normal-hearing (diamonds) and four sensorineural hearing-impaired subjects (boxes) are shown. The left panel represents the $\Delta f = 200$-Hz condition and the right panel represents the $\Delta f = 16$-Hz condition. Running Gaussian noise at a center frequency of 2 kHz was used. Carrier and modulation duration: 1 s. Levels: (see Tab. 2.1). (For reasons of visibility, the il-data are slightly shifted to the right).

four sensorineural hearing-impaired listeners. For both groups, no systematic influence of loudness on modulation-detection thresholds can be seen both for the Δf = 200-Hz or the Δf = 16-Hz condition. For each group and condition the thresholds across subjects vary within a range of 3.5 dB. The thresholds for the hearing-impaired listeners are slightly but significantly (Wilcoxon Rang Sum Test, $p < 0.05$) higher than for the normal-hearing listeners.

2.2.3 Simulations

Figure 2.5 shows predicted (filled symbols) and mean measured (open symbols) modulation detection thresholds as function of modulation frequency for the carrier bandwidth of Δf = 200 Hz (left panel) and for the carrier bandwidth of Δf = 16 Hz (right panel). The simulations were obtained with the modulation-filterbank model for two presentation levels according to the mean presentation level for the normal-hearing listeners (60 dB SPL, filled diamonds) and for the hearing-impaired listeners (90 dB SPL, filled boxes). Each point represents the mean value of 20 simulated experimental runs.

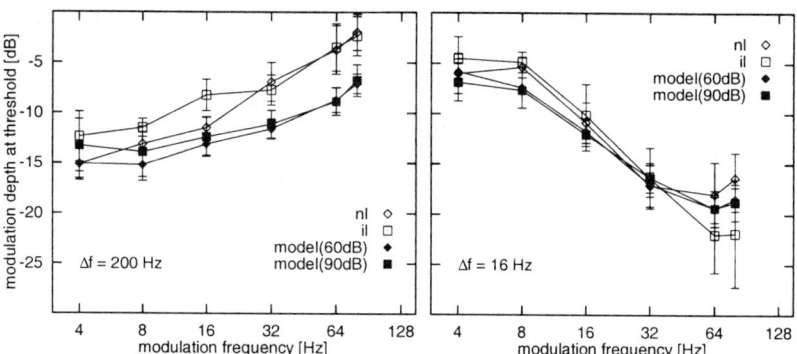

Fig. 2.5: Measured (open symbols) and simulated (filled symbols) modulation-detection thresholds as function of modulation frequency. The measured data represent mean thresholds of five normal-hearing (open diamonds) and of four sensorineural hearing-impaired subjects (open boxes). Model predictions were performed at 60 (filled diamonds) and 90 dB SPL (filled boxes) corresponding to the mean presentation level for normal-hearing and hearing-impaired listeners, respectively. The left panel represents the Δf = 200-Hz bandwidth condition and the right panel represents the Δf = 16-Hz bandwidth condition. Running Gaussian noise at a center frequency of 2 kHz was used. Carrier and modulation duration: 1 s. Level: corresponding to 25 CU (see Tab. 2.1).

2.2 Experiment I: Modulation detection with stochastic carriers

The left panel shows the averaged mean data from Fig. 2.2 for both normal-hearing (open diamonds) and hearing-impaired listeners (open boxes) for the carrier bandwidth of $\Delta f = 200$ Hz. The shapes of the threshold curves and the absolute values are very similar for both groups. At modulation rates below 16 Hz however, the threshold-values are about 2 dB higher for the hearing-impaired listeners. Only in the case of the 16 Hz modulation the standard deviations for the normal-hearing and hearing-impaired listeners do not overlap. The predicted threshold curves show increasing thresholds with increasing modulation frequency (thus reflecting a lowpass-characteristic of the TMTF). The slope of the simulated thresholds agrees well with the experimental data and is also consistent with similar model predictions presented by Dau et al. (1997a). The simulated thresholds underestimate the amount of masking for high modulation frequencies by about 5 dB.

The right panel of Fig. 2.5 shows the corresponding data for the $\Delta f = 16$-Hz condition (averaged data from Fig. 2.3). For modulation frequencies up to $\Delta f / 2$, the thresholds remain constant and decrease for higher modulation frequencies up to 64 Hz (thus reflecting a highpass-characteristic of the TMTF). The simulated thresholds agree well with the experimental data and are also consistent with similar model predictions presented by Dau et al. (1997a). Except for the modulation frequencies of 64 and 80 Hz, the thresholds are identical for both subject groups. Note however, that due to the large interindividual differences at these frequencies the average across subjects shows a large variability so that this deviation should not be considered further.

2.3 Experiment II: Intensity discrimination and modulation detection with sinusoidal carriers

2.3.1 Method

2.3.1.1 Procedure and subjects

The procedure was the same as described in Exp. I (Sec. 2.2), except that the subjects received visual feedback during the experimental runs. Three male normal-hearing subjects (audiometric thresholds within 10 dB HL) participated voluntarily in the experiment. All subjects were trained in psychoacoustic measurements. They were aged from 28 to 33 years. Three male and one female sensorineural hearing-impaired subjects participated in the experiment. Two of them had experience in psychoacoustic measurements but none of them participated in the experiment described in Sec. 2.2. They were aged from 31 to 73 years. Figure 2.6 shows the audiometric thresholds of the hearing-impaired listeners. In the region around 5 kHz, all hearing-impaired subjects show a flat (il1, il2, il3) or a mild downward-sloping (il4)

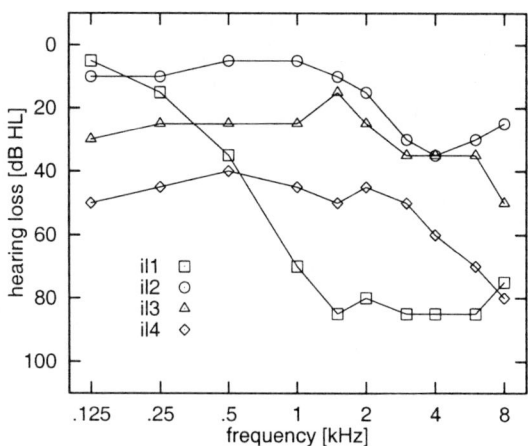

Fig 2.6: Audiometric thresholds for four sensorineural hearing-impaired listeners participating in Exp. II. The ordinate indicates the hearing loss in dB HL and the abscissa represents the frequency of the tone.

2.3 Experiment II: Intensity discrimination and modulation detection ...

hearing loss between 35 and 85 dB HL. The threshold difference between bone- and air-conduction was less than 10 dB in the specified region. In case of an asymmetric hearing loss, always the better ear was tested. The hearing-impaired listeners were paid for their participation on an hourly basis.

2.3.1.2 Apparatus and stimuli

All acoustic stimuli were digitally generated at a sampling frequency of 32 kHz. In contrast to Exp. I (Sec. 2.2), the on-board 16-bit D/A converters of a SGI INDY work station were used to transform digital to analog stimuli. The other equipment was the same as in Exp. I.

In the modulation-detection task, the test signal was a sinusoidally amplitude modulated (SAM) 5 kHz tone of 1 s duration. The modulation frequency to be detected was 16 Hz. The modulation had a duration of 750 ms and was temporally centered in the carrier. A $\cos^2$-shaped window was applied over the entire duration of the modulation. In addition, 125 ms-$\cos^2$-shaped ramps were applied to the reference and the test stimulus. Finally, each interval was adjusted to have the same rms-value. In the intensity-discrimination task, the level of the reference tone was increased in the test-signal interval. The increment had the same duration as the modulation in the previous experimental condition. It was also centered temporally in the 5 kHz tone and a $\cos^2$-shaped window was applied over the entire duration of the increment. This experimental condition may also be interpreted as a 0-Hz modulation and will be referred to as „0-Hz condition".

The left panel of Fig. 2.7 shows the envelope of the reference (dashed line) and of the test signal (solid line) for the 16-Hz condition. The right panel shows both envelopes for the 0-Hz condition. The experiment was performed at two presentation levels which equaled the perceived loudness categories „mittellaut" (CU 25, i.e., „intermediate") and „sehr leise" (CU 5, i.e., „very soft"). The presentation level was set to 60 dB SPL and 20 dB SPL, respectively, for the normal-hearing listeners and was chosen individually for the hearing-impaired listeners (cf. Tab. 2.2).

categories [CU]	listener mean	nl	il1	il2	il3	il4
25 „mittellaut" („intermediate")		60	105	60	80	75
5 „sehr leise" ("very soft")		20	80	45	45	65

Tab. 2.2: Presentation levels in dB SPL for the normal-hearing and the hearing-impaired listeners. The categorical units correspond to the perceived loudness given in quotes.

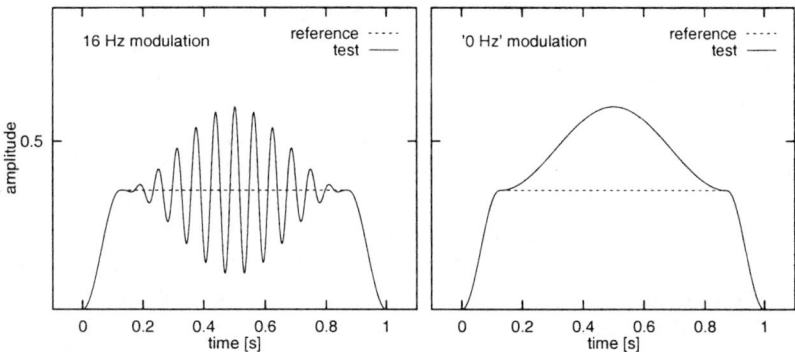

Fig. 2.7: Envelope of the reference stimulus (dashed line) and the test stimulus (solid line) for the 16-Hz condition (left panel) and the 0-Hz condition (right panel). Carrier: 5-kHz tone.

2.3.2 Results

2.3.2.1 Increment- and modulation-detection thresholds

In this experiment, the intensity-discrimination (0-Hz condition) thresholds for the 5-kHz tone, and the 16 Hz modulation-detection thresholds imposed on a 5-kHz carrier were measured. Fig. 2.8 shows the experimental results as median values (interquartiles lower than the symbol size) of four experimental runs for each subject. The upper panels of Fig. 2.8 show results obtained at a presentation level corresponding to 25 CU and the lower panels show results obtained at a presentation level corresponding to 5 CU (see Tab. 2.2). The ordinate indicates modulation depth at threshold, and the abscissa represents the modulation frequency. The left panels of Fig. 2.8 show the thresholds for three normal-hearing subjects and the mean threshold-values as dotted lines. The right panels of Fig. 2.8 show the thresholds for four sensorineural hearing-impaired subjects.

2.3 Experiment II: Intensity discrimination and modulation detection ... 31

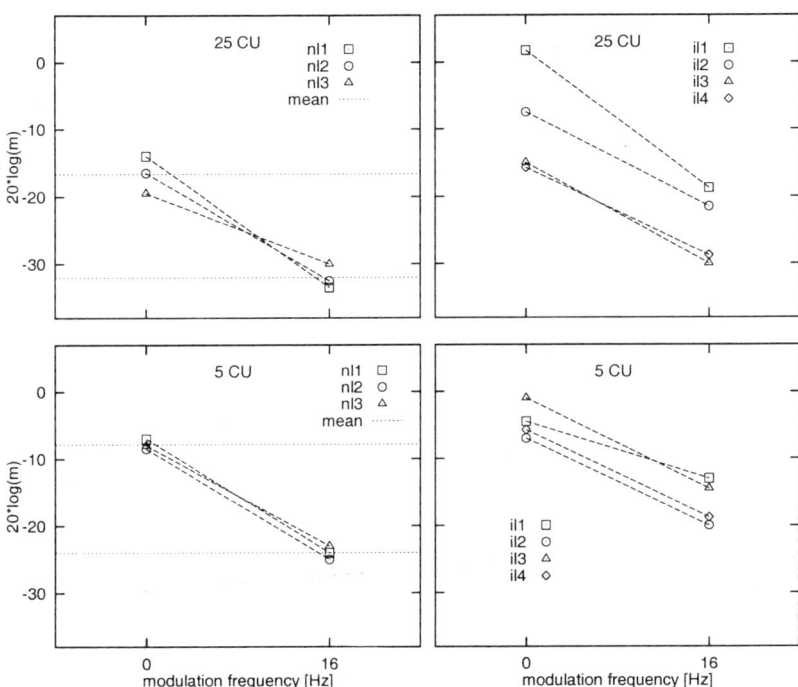

Fig 2.8: Increment-detection thresholds for a 5 kHz tone and modulation-detection thresholds for a 16-Hz modulation imposed on a 5 kHz tone for five normal-hearing listeners (left panels) and four sensorineural hearing-impaired listeners (right panels). The ordinate indicates the detection threshold in dB and the abscissa represents the modulation frequency. The presentation level corresponds to 25 CU (upper panels) and 5 CU (lower panels) (see. Tab. 2.2).

For the higher presentation level (upper left panel), the data of the normal-hearing subjects indicate a mean threshold of -17 dB for the 0-Hz condition which is slightly larger than the typical 1-dB criterion which corresponds to -18.27 dB[1]. The mean threshold for the 16-Hz condition is -32 dB which is 15 dB lower. In contrast to the data for the normal-hearing listeners, corresponding data of the hearing-impaired subjects (upper right panel) show large interindividual differences. The hearing-impaired listeners il3 and il4 show very similar thresholds as the normal-hearing listeners

---

[1] A 1-dB level increment is achieved if the signal amplitude is multiplied by the factor 1.122. This can be expressed as a modulation: $s(t) = c(t)(1+m)$, with the modulation depth of $m=0.122$. $(20\log(m) = -18.27 \text{ dB})$.

while il2 and il1 show much higher thresholds in both conditions tested (0-Hz and 16-Hz condition).

The mean thresholds for the normal-hearing listeners obtained at a presentation level corresponding to 5 CU (lower left panel) are about 9 dB higher in both conditions compared to the results obtained at the higher presentation level of 25 CU (upper left panel). This is consistent with data from the literature (e.g., Kohlrausch, 1993). The mean threshold of -8 dB corresponds to a 2.9-dB increment in intensity. The mean threshold for the 16-Hz condition is again about 15 dB lower than the threshold for the intensity increment. In general, the thresholds for the hearing-impaired listeners are higher than the thresholds for the normal-hearing listeners. As for the higher presentation level, interindividual deviations for the hearing-impaired listeners (lower right panel) are larger than those for the normal-hearing listeners (lower left panel). Note that listener il2 shows markedly increased thresholds compared to the results of the normal-hearing listeners at the high presentation level (upper panels) while at the low presentation level, thresholds are much more similar to those for the normal-hearing listeners (lower panels). For listener il3, on the other hand, thresholds are near normal at the high presentation level but are markedly increased at the low presentation level.

Table 2.3 shows mean threshold differences (and interindividual standard deviations) between the 0-Hz and the 16-Hz condition. At the high presentation level, the mean threshold differences and the standard deviations are similar for both normal-hearing listeners and hearing-impaired listeners. At a level corresponding to 5 CU, the mean threshold differences are about 4 dB lower for the group with sensorineural hearing impairment than for the group with normal hearing.

listener categories [CU]	nl mean ± σ	il mean ± σ	model
25 „mittellaut" („intermediate")	15.3±4.5	15.6±3.4	11
5 „sehr leise" ("very soft")	16.2±1.0	12.0±2.3	10

Tab. 2.3: Threshold differences between the 0-Hz and the 16-Hz condition. Mean differences and interindividual standard deviations for both groups of listeners and corresponding model predictions are given in dB.

2.3.3 Simulations

Figure 2.9 shows predicted thresholds for the 0-Hz (solid lines) and the 16-Hz (dashed lines) condition as a function of the variance of the internal noise assumed in the model. Simulations were performed at two presentation levels: 60 dB corresponding to 25 CU (left panel) and 20 dB corresponding to 5 CU (right panel). The symbols (circles, triangles) represent the thresholds for a normalized variance of 1. The circle in the left panel represents the standard intensity-discrimination condition to which the model initially has been adjusted to approximately satisfy Weber's law, i.e. the 1-dB criterion (Dau et al. 1996a, 1997a). It is assumed within the model that the internal noise limits the resolution as long as deterministic stimuli are used. For the 16-Hz condition (triangle) the predicted threshold is -29 dB, i.e., 11 dB lower than in the 0-Hz condition (-18 dB). This is 4 dB less than the measured threshold difference of 15 dB. For the low presentation level (right panel) both predicted thresholds are about 6 dB higher than in the higher level condition, while the threshold difference remains nearly constant (10 dB). Such a shift (9 dB) of both thresholds to higher values is also visible in the experimental data.

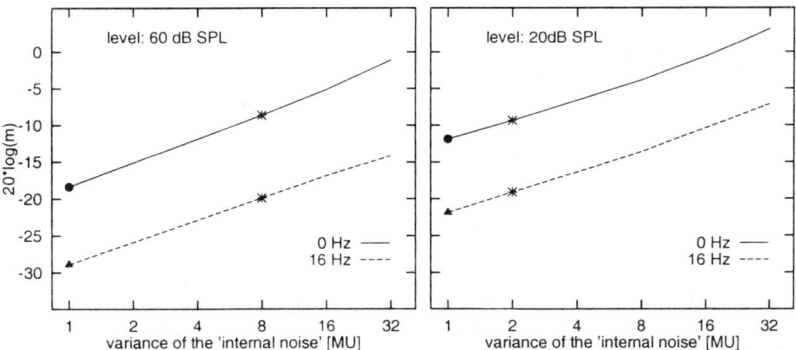

Fig 2.9: Simulated detection thresholds for the 0-Hz condition (solid lines, circles) and the 16-Hz condition (dashed lines, triangles) as function of the variance of the internal noise. The left panel indicates the thresholds at a simulated presentation level of 60 dB SPL (25 CU). The right panel at a presentation level of 20 dB SPL (5 CU). The filled symbols indicate the thresholds for the standard model (normalized variance of 1). The stars indicate predicted thresholds for an individual hearing-impaired listener (il2).

A doubling of the variance of the internal noise leads to 3 dB higher thresholds, independent of the simulated condition and the presentation level, i.e., the predicted threshold difference between the 0-Hz and the 16-Hz condition remains constant. This allows to simulate the measured thresholds for an individual hearing-impaired listener by adjusting the variance of the internal noise. For example, the thresholds for listener i12 at the high presentation level (see Fig 2.8, upper right panel) would need the eight-fold variance, leading to an intensity discrimination threshold of -8.5 dB and a modulation detection threshold of -20 dB (see Fig. 2.9, stars). However, for the low presentation level only about the two-fold variance would be appropriate for simulating the thresholds for listener i12 (see Fig. 2.8, lower right panel).

2.4 Discussion

2.4.1 Stochastic carrier

The results from Exp. I (Sec. 2.2) indicate that modulation detection thresholds obtained with stochastic narrowband noise carriers are *not* affected by the amount of hearing loss of the sensorineural hearing-impaired listeners. The shape of the TMTF as well as the absolute values of the obtained thresholds are the same as for the normal-hearing listeners. Hence, the hearing-impaired listener's ability to process temporal envelope fluctuations and the frequency selectivity for modulations appear to be the same as that of the normal-hearing listeners. This is at least the case when the stimuli are presented at the same perceived loudness.

It is widely accepted in the literature (e.g. Moore, 1995) that a loss of compression in basilar membrane velocity is a main factor in sensorineural hearing impairment. This loss causes the recruitment phenomenon (reduced auditory dynamic range) which is present in all four hearing-impaired subjects from Exp. I (cf. Tab. 2.1). Our modulation detection data therefore provides evidence that peripheral compression does not affect the ability of the auditory system to detect amplitude modulations. If one assumes that the signal-to-noise ratio of test and masker modulation at a certain processing stage determines the threshold, this behavior can be explained as follows: The compression (or the loss of it) affects both the inherent modulations of the carrier itself and the imposed test modulation in the same way. Consequently, the same detection threshold can be expected in the conditions with stochastic carriers considered here.

As described in Dau et al. (1997a+b), the threshold for an amplitude modulation is primarily determined by the modulation masking effect of the inherent fluctuations of the carrier itself and the auditory system's frequency selectivity for modulation. More specifically, the integrated modulation power of the carrier within the transfer range of the modulation filter tuned to the actual modulation frequency already gives a very good estimate for the detection threshold ("modulation filterbank model", Dau et al., 1997a, Dau et al., 1999). The model's ability to simulate the shape and the absolute values of the threshold pattern for the different bandwidth conditions results primarily from the proposed frequency selectivity for modulations in the respective modulation filter. As a consequence of the data and model predictions given here, the same modulation filter parameters should be assumed for sensorineural hearing-impaired listeners as in the model for normal-hearing listeners.

2.4.2 Deterministic conditions

In Exp. II (Sec 2.3), deterministic stimuli were used which do not contain any statistical envelope fluctuation so that only internal noise limits the resolution of the auditory system. Intensity-discrimination thresholds were compared with modulation-detection thresholds. The modulation depth m at the modulation detection threshold, is about 15 dB below the value for m that corresponds to the 1-dB criterion in the intensity-discrimination task. This is the case for both normal-hearing and hearing-impaired listeners. The question is how this large difference of 15 dB can be explained. A simple energy detector which would predict the 1-dB criterion in the intensity-discrimination task, fails in the modulation-detection task since the overall energy remains the same (e.g., Viemeister, 1979). A max.-to-min. criterion, as supposed by Forrest and Green (1987) for describing modulation detection thresholds, would predict a 6-dB difference in threshold for the two tested conditions, but never a 15 dB effect. In order to account for this difference in the experimental conditions one has to assume either an „asymmetry" in the decision criterion or some asymmetry in the signal processing.

Within the present model, such an asymmetry in the signal processing is provided by the adaptation stage, which analyzes fluctuating stimuli in a different way as stationary stimuli: While stationary signals are compressed in an approximately logarithmic way, fluctuating stimuli are transmitted with less compression. This reduced compressive property for modulations results in the high sensitivity for modulation detection in the model. Hence, it is responsible for the predicted threshold difference of 11 dB between the two experimental conditions under consideration. Since all sensorineural hearing-impaired subjects, independent of their hearing loss, show essentially the same difference in thresholds as the normal-hearing subjects, it appears, that some retrocochlear compression effectively occurs in the auditory processing of stationary stimuli in both groups of subjects.

In contrast to the threshold difference, the *absolute* thresholds for both, the intensity-discrimination and the modulation-detection task are markedly increased in some of our hearing-impaired listeners. Note that this

2.4 Discussion

increase in the threshold values is not determined by the amount of hearing loss (compare Fig. 2.6 and Fig. 2.8). This indicates that the increase in threshold values is due to a general loss of sensitivity affecting both intensity discrimination and modulation detection. In the model, such a general loss of sensitivity can be modeled by an increase of the variance of the internal noise.

Taken together, for the modeling of an individual sensorineural hearing impairment, at least two modifications have to be introduced within the model. The variance of the internal noise has to be increased to account for a general loss of sensitivity. The compressive properties of the model have to be reduced to account for an altered loudness scaling. It is also likely, that a reduced spectral resolution is necessary for simulating impaired hearing. However, these modifications are *not* needed to account for the shape of the TMTF, i.e., the temporal acuity of the impaired auditory system.

2.5 Conclusions

- The experiments on modulation detection with stochastic carriers demonstrate that the processing of amplitude modulations is the same for normal-hearing and sensorineural hearing-impaired listeners tested in this study. A loss of peripheral (cochlear) compression does not affect the ability of the auditory system to detect modulations.

- The experiments on intensity discrimination and modulation detection with a sinusoidal carrier show similar results for normal-hearing and hearing-impaired listeners. For some hearing-impaired subjects, a general loss of sensitivity in both experimental conditions is observed. This general loss is not directly related to the amount of sensorineural hearing impairment.

- The proposed processing model is capable of quantitatively modeling most aspects of the modulation-masking, modulation-detection and intensity-discrimination experiments described. The large difference in thresholds for intensity discrimination and modulation detection, respectively, is accounted for by the adaptation stage of the model: It performs a retrocochlear compression for stationary stimuli whereas modulations are transformed in a less compressive way.

- A model for describing the signal processing of sensorineural hearing-impaired subjects should employ the same temporal processing stages and frequency-selectivity stage for modulations as for normal-hearing listeners. An increase of the variance of the internal noise would account for the observed general loss of sensitivity in some hearing-impaired listeners.

Chapter 3
Modeling loudness matching and modulation matching

Abstract

Modulation matching experiments with unilateral sensorineural hearing-impaired listeners (Moore et al., 1996) showed that a given modulation depth in the impaired ear was matched by a greater modulation depth in the normal ear. The modulation-matching functions could be predicted reasonably well by the results from a loudness-matching experiment. This was shown for modulation rates in the range from 4-32 Hz. The present study shows results from corresponding simulations based on an auditory detection model for normal-hearing listeners (Dau et al., 1996a+b, 1997a+b) which has been modified for hearing-impaired listeners. The modifications of this model include an elevated absolute threshold and a less compressive characteristic within the adaptation stage of the model. The detection device is replaced by a simple pattern comparison device at the level of the internal representation of the stimuli. The study is focused on the question whether the processing of amplitude modulations in hearing-impaired listeners can be described within the model by a modulation-rate-dependent or -independent reduction in compression. The model predictions obtained with an instantaneous, modulation-rate-independent expansion appear more consistent with the data than the modulation-rate-dependent version realized by reducing the number of nonlinear adaptation loops. For a final decision, however, data obtained with modulation rates larger than 32 Hz are required.

3.1 Introduction

Differences in the auditory signal processing between normal-hearing listeners and hearing-impaired listeners become most easily apparent if the sensation „loudness", produced by a given stimulus, is compared. Most sensorineural hearing-impaired listeners show a phenomenon referred to as loudness recruitment (Fowler, 1936, Steinberg and Gardner, 1937). That is, the absolute threshold for a given stimulus is elevated, but the stimulus level which is uncomfortable or causes pain to the listener remains nearly unchanged compared to that of normal-hearing listeners[1]. Thus, the dynamic range of signal levels usable for signal perception is reduced for sensorineural hearing-impaired listeners. The mechanisms underlying loudness recruitment are not fully understood. However, the general accepted assumption is that it results primarily from the loss of the compressive nonlinearity that occurs in the normal peripheral auditory system (Yates, 1990, Zeng and Turner, 1991, Moore, 1995). The compressive nonlinearity in a normal ear seems to operate very quickly. The compression can be observed in physiological studies using tone bursts lasting only a few tens of milliseconds (Ruggero, 1992) or using click-stimuli where a nearly instantaneous compression is observed (Recio et al., 1998). In psychoacoustical studies, effects associated with peripheral compression were observed with stimuli of 10 ms duration in masking experiments (Carlyon and Datta, 1997, Oxenham and Plack, 1997, Plack and Oxenham, 1998).

Most studies of loudness recruitment have used steady sounds of relatively long duration, such as tone bursts or bursts of noise (Moore et al., 1985, Pluvinage, 1989, Allen and Jeng, 1990, Kissling et al. ,1993, Kollmeier and Hohmann, 1995). In a study by Moore et al. (1996) the effects of loudness recruitment on steady sounds and dynamically-varying sounds were examined by loudness-matching and modulation-matching experiments with unilateral sensorineural hearing-impaired listeners. Thus, a direct comparison between a normal and a recruiting ear was performed for steady sounds and for temporally-varying sounds. It was shown that a given modulation depth in the impaired ear was matched by a greater modulation

---

1 There are several different kinds of recruitment observed with sensorineural hearing-impaired listeners. For some hearing-impaired subjects, the UCL (uncomfortable loudness level) is reached either at lower (over-recruitment) or at higher (under-recruitment) stimulus levels than for normal-hearing listeners. The most common pattern is the complete recruitment where the UCL is about equal to that of normal-hearing listeners. Complete recruitment is considered in this study.

3.1 Introduction

depth in the normal ear for modulation rates in the range from 4-32 Hz. The modulation-matching functions could be predicted reasonably well by the loudness-matching results obtained with steady tones. These results are in line with the idea that a main factor of loudness recruitment is the loss of fast-acting compressive nonlinearity that operates in the normal peripheral auditory system.

This paper concentrates on the question how such a loss of peripheral compression can be understood in terms of the processing model by Dau et al. (1996a+b, 1997a+b). This model was selected since it quantitatively describes a large variety of psychoacoustical effects in normal-hearing listeners including forward-masking, temporal integration, modulation perception, spectral masking (Derleth et al., 1999) and some aspects of speech perception in normal and hearing-impaired listeners (Holube and Kollmeier, 1996). The required modifications of the model to account for the data are of general interest, since they clarify the differences between a normal and an impaired auditory system in conditions of both stationary and time-varying sounds. The processing model was originally designed to describe simultaneous and nonsimultaneous masking data for normal-hearing listeners. This was achieved by a stage which combines temporal adaptation effects with compressive properties. These compressive properties are time dependent. Within the model the adaptation stage is realized by a chain of five nonlinear adaptation loops (Püschel, 1988) with time constants in the range from 5 to 500 ms. The two loops with the shortest time constants (5 and 50 ms) are assumed to represent the compressive properties of the peripheral auditory system. One approach to introduce a loss of peripheral compression within the model for hearing-impaired listeners would be to skip the adaptation loops with the shortest time constants. Such an approach would also affect temporal properties of the model. A different approach, which would leave the temporal properties of the model unchanged, realizes a loss of compression by an instantaneous expansion prior to the chain of the original adaptation loops. To decide which model approach is more adequate for simulating sensorineural hearing impairment, it is necessary to consider experiments with steady sounds and with temporally-varying sounds which both are assumed to be affected by the loss of peripheral compression. Therefore, simulations of the loudness matching (Sec. 3.3) and modulation matching data (Sec. 3.4) by Moore et al. (1996) obtained with unilateral sensorineural hearing-impaired listeners are presented.

3.2 Model

Simulations were performed with the modulation-filterbank model by Dau et al. (1997a+b). The original model consists of several preprocessing stages and an optimal detector as the decision device. The first processing stage is the linear Gammatone-filterbank model suggested by Patterson et al. (1987). In the simulations of the present study, only the peripheral channel tuned to the signal frequency is considered. After peripheral filtering, the model contains half-wave rectification, lowpass filtering at 1 kHz and an adaptation stage. The adapted signal is then processed by a modulation filterbank (Dau et al. 1997a+b). An internal noise with a constant variance is added to the output of each modulation filter to limit the resolution within the model. The signal representation at this stage of the model is referred to as its „internal representation". For the simulation of detection experiments, the optimal detector compares the internal representation of the current signal with a suprathreshold template representation of the signal. For the simulation of sensation qualities, however, alternative pattern processing algorithms can be employed.

One possible interpretation of the absolute value of the stimulus' internal representation is that it may represent a correlate of perceived loudness when integrated across frequency bands. Hence, the internal representation of the same stimulus should be different for normal-hearing listeners and for hearing-impaired listeners reflecting their differences in loudness perception. This has to be achieved by an altered (less compressive) adaptation stage in the model for impaired hearing. To simulate the sensation of loudness, simply the mean-value of the internal representation, calculated across a duration of 100 ms, is associated with loudness categories of a categorical loudness-scaling experiment (Heller, 1985, Hohmann and Kollmeier, 1995). The mean excitation is expressed in model units (MU). Consequently, they are associated with verbal categories: ≤ 0 „inaudible", 10 „very soft", 30 „soft", 50 „intermediate", 70 „loud", 90 „very loud" and ≥ 100 „too loud".

For the simulation of other sensation qualities than the perceived loudness, different ways to process the „internal representation" have to be assumed within the model. In Sec. 3.4, model predictions concerning the sensation of modulation depth are presented. They are derived within the

3.2.1 Adaptation stage for normal hearing

Figure 3.1 shows the adaptation stage of the model for normal hearing **nh** (left panel) as proposed by Dau et al. (1996a+b, 1997a+b). Originally, they have been developed to represent compressive and adaptive properties of the intact auditory system. This was realized by a chain of consecutive nonlinear adaptation loops. The initial part of the adaptation stage is a limiter which restricts the input prior to the adaptation loops to a minimal value (*min*) to establish an absolute threshold for auditory stimuli. This minimum value corresponds to the expectation value of the envelope of the peripheral internal noise. A conductive hearing loss can easily be simulated by an increase of the value *min*. The second part of the adaptation stage consists of a chain of five consecutive nonlinear adaptation loops (Püschel, 1988) with different time constants τ_n. Each adaptation loop consists of a divider and a lowpass filter. The lowpass-filtered output of the adaptation loop is fed back to form the denominator of the dividing element. The divisor is the momentary charging state of the lowpass filter. The time constants are set to values of $\tau_1=5$, $\tau_2=50$, $\tau_3=129$, $\tau_4=253$ and $\tau_5=500$ ms. For stationary stimuli, an input value I produces a value $O = I^{1/2}$ at the output of the first adaptation loop (derived from the stationary condition $I / O = O$). For a chain of n loops we obtain an output of:

$$O = I^{(\frac{1}{2})^n} \qquad (3.1)$$

For n = 5 this approaches a logarithmic transform if the output values for stationary inputs with levels between 0 and 100 dB are mapped[1] to the range of 0 to 100 model units (MU). Input variations that are rapid compared to the time constants of the lowpass filters are transformed in a less compressive way than stationary stimuli. In chapter two it was shown that this different processing of stationary and fluctuating stimuli enables the model to account both for intensity discrimination and modulation detection.

---

[1] The input range of 100 dB is represented within the model by values between 10^{-5} (*min*) and 1. The mapping of the output-values of the last adaptation loop to the range between 0 and 100 model units (MU) is achieved by subtracting the minimal possible output-value for a stationary stimulus ($min^{1/32}$) and by multiplication with the factor ($100 / (1 - min^{1/32})$).

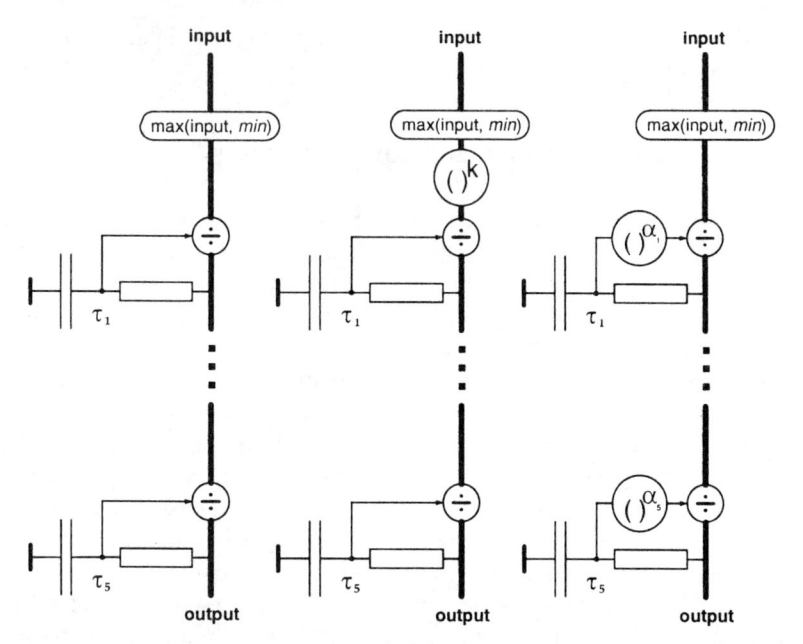

Fig. 3.1: Schematic plot of the adaptation stage of the model for normal hearing **nh** (left panel) and for the two model versions for impaired hearing **ih1** (middle panel) and **ih2** (right panel) that work on the bandpass-filtered, halfwave-rectified and lowpass-filtered auditory stimulus. The original set of five consecutive nonlinear adaptation loops (Püschel, 1988) for modeling temporal adaptation effects and compressive properties of the normal auditory system are used in the models **nh** and **ih1**. In model **ih1** the stimulus envelope is raised to the power of k (k ≥ 1) prior to the adaptation loops. The adaptation stage of model **ih2** consists of five modified adaptation loops. The compressive properties of each loop can be adjusted by the value of the exponent α_n ($\alpha_n \leq 1$).

3.2.2 Adaptation stage for sensorineural impaired hearing

For modeling sensorineural hearing impairment, the „effective" compression of the model has to be reduced. Figure 3.1 shows the adaptation stage of two model versions for impaired hearing, **ih1** (middle panel) and **ih2** (right panel), respectively. Model **ih1** realizes a reduced compression by an instantaneous, i.e. sample-by-sample, expansion prior to the adaptation stage for normal hearing. Thus, the dynamic properties of the adaptation stage remain unchanged. Model **ih2** realizes a reduced compression of the adaptation stage by reducing the compressive properties of each adaptation

3.2 Model

loop. Such a processing leads to the same results with stationary signals as **ih1**, but differs in the transformation characteristic for time-varying stimuli.

For the adaptation stage of the model **ih1** for impaired hearing (middle panel of Fig. 3.1), each input sample is taken to the power of k (k ≥ 1) prior to the processing by the adaptation loops (see above). A stationary input I to this adaptation stage is transformed to an output O of:

$$O = I^{k(\frac{1}{2})^5} = I^{\frac{k}{32}} \tag{3.2}$$

For k = 1, this adaptation stage is equivalent to the adaptation stage for normal hearing. The subsequent mapping to model units is exactly the same as in the model for normal hearing. Therefore, the input range of 0 to 100 dB to the adaptation stage is mapped to a larger range of model units. Depending on the value of k, stimuli up to a certain level are mapped to negative model units. Such stimuli are assumed to be „inaudible". An input level corresponding to 100 dB SPL is mapped to 100 MU as in the model for normal hearing, i.e., such a signal is assumed to be „too loud" within both models, **nh** and **ih1**.

Within the model **ih2** (right panel of Fig. 3.1), a reduced compression for stationary stimuli is achieved by a reduced compression within each adaptation loop. This is achieved by taking the momentary charging state of the lowpass to the power of α_n ($\alpha_n \leq 1$) before it enters the divisor. The input value I to such a modified adaptation loop is transformed to an output-value O in the stationary condition according to $O = I / O^{\alpha}$. For the complete set of five adaptation loops the transformation amounts to:

$$O = I^{\prod_{n=1}^{5} \frac{1}{(1+\alpha_n)}} \tag{3.3}$$

For $\alpha_{1-5} = 1$, this adaptation stage is equivalent to the adaptation stage for normal hearing. As before for model **ih1**, the subsequent mapping to model units is the same as in the model for normal hearing. Thus, the input range of 0 to 100 dB to the adaptation stage is mapped to a larger range of model units. Depending on the values of the α_n, stimuli up to a certain level are mapped to negative model units (corresponding to inaudible sensations), while an input value corresponding to 100 dB SPL is again mapped to 100 MU.

3.3 Loudness matching for sinusoidal stimuli

3.3.1 Method

3.3.1.1 Experimental procedure and stimuli

In a study by Moore et al. (1996), loudness matches between the normal and the impaired ear were obtained for three unilateral sensorineural hearing-impaired listeners. A 1-kHz sinusoid gated with a 200-ms-steady-state portion and 10-ms-raised-cosine rise and fall was used. The stimuli were presented in regular alternation to the two ears with 500 ms interstimulus intervals. Within a run, the tone was fixed in level in one ear (either the normal or the impaired ear) and the level in the other ear was varied to determine the level corresponding to equal loudness. The range of levels was between 20 dB SL and 90 dB SPL in the normal ear and between 10 dB SL and 100 dB SPL in the impaired ear. An adaptive procedure with a minimum step size of 1 dB was used to obtain five matches for each stimulus level tested. The present study employs the data of the study by Moore et al. (1996), averaged across subjects for a comparison with model predictions.

3.3.1.2 Simulations

Simulations were performed with the model version for normal hearing **nh** and both model versions for simulating impaired hearing **ih1** and **ih2**. The input/output characteristic for each model version was determined separately for the stimuli employed by Moore et al. (1996). Therefore, the mean value of the internal representation between 110 and 210 ms was calculated for each stimulus level in the range between 0 and 100 dB SPL (step-size: 1 dB). Loudness matching functions were obtained between the model versions **nh-ih1** and **nh-ih2**. That pair of input levels which lead to the same output-value in MU for both model versions, i.e., to the same „perceived" loudness, are taken as loudness-match. Matches were obtained for loudness categories corresponding to 0, 10, 30, 50, 70, 90 and 100 MU. The parameter $k = 3$ was chosen for the expansion in model **ih1**. In model **ih2**, the parameters α_1, α_2 and α_{3-5} were set[1] to 0, 0.4 and 1, respectively. These

---

[1] The reason for changing the compressive properties of the adaptation loops with the shortest time constants was, that it is assumed that the peripheral compression on the basilar membrane is represented by the most fast-acting adaptation loops.

3.3 Loudness matching for sinusoidal stimuli

values were found to match the average empirical data from the three subjects reported by Moore et al. (1996). Note, however, that no explicit procedure was employed for an optimum parameter fit to the data.

3.3.2 Results

The left panel of Fig. 3.2 shows the input/output characteristics obtained with model **nh** (triangles), **ih1** (squares) and **ih2** (circles) for a 1 kHz tone. The abscissa indicates the stimulus level in dB SPL. The ordinate represents model units which are associated with loudness categories. The input/output function for model **nh** is close to a straight line with slope 1, reflecting a logarithmic compression. The nearly identical input/output functions for models **ih1** and **ih2** consist of a constant part at 0 MU up to an input-level of 65 dB and an increasing part with a nearly constant slope of 2.9 MU/dB for higher input-levels. Thus, the model **nh** simulates an absolute threshold at 0 dB and an usable dynamic range of 100 dB, while the models **ih1** and **ih2** simulate an increased absolute threshold (hearing loss) at 65 dB and a remaining usable dynamic range of 35 dB.

The right panel of Fig. 3.2 shows the mean loudness-matching function for three unilateral sensorineural hearing-impaired subjects (open diamonds, replot from Moore et al., 1996) together with model predictions (filled diamonds). The respective lowest symbol indicates the absolute thresholds both for the normal and the impaired auditory system. The highest symbols indicate that stimulus level which produces an equal-loud sensation for the normal and the impaired auditory system. The horizontal and the vertical lines indicate one standard deviation for the impaired and for the normal ear, respectively, derived from the experimental data. The dashed part of the curve indicates the range of levels for which loudness matches were measured. The predicted loudness-matching function (filled diamonds) results from the combination of the input/output functions for the different model versions **nh-ih1** and **nh-ih2** (left panel of Fig. 3.2). Because of the similar input/output functions of both model predictions for impaired hearing, the predicted loudness-matching function represents both model-combinations.

Chapter 3: Modeling loudness matching and modulation matching

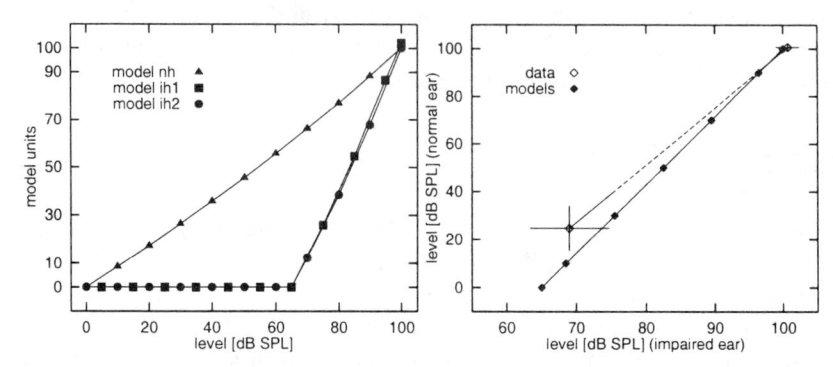

Fig. 3.2: The left panel shows input-output functions for the model **nh** (triangles), **ih1** (squares) and **ih2** (circles). The input-stimulus was a 1 kHz tone. Model units are associated with loudness-categories (≤ 0 „inaudible", 10 „very soft", 30 „soft", 50 „intermediate", 70 „loud", 90 „very loud" and ≥ 100 „too loud"). The right panel shows loudness-matching functions for the same stimulus. Mean data from Moore et al. (1996) for three unilateral sensorineural hearing-impaired listeners are indicated by open diamonds. The errorbars indicate one standard deviation for the impaired ears (horizontal) and for the normal ears (vertical). The dashed part of the curve indicates the range of stimulus levels for which loudness-matches were measured. Corresponding model predictions (for both model combinations **nh-ih1** and **nh-ih2**) are indicated by the filled symbols. The lowest diamond indicates the absolute thresholds. The highest diamond indicates that stimulus level which produces an equal-loud sensation in the normal and in the impaired auditory system.

The predicted stimulus level which produces an equal-loud sensation (highest filled diamond) is in line with the data. The predicted absolute threshold (lowest filled diamond) for the normal auditory system is markedly (23 dB) lower than the measured absolute threshold for the so called „normal" ear of the unilateral hearing-impaired listeners. Therefore, the predicted loudness-matching function is slightly steeper (2.9 dB/dB) than the measured loudness-matching function (2.4 dB/dB). Nevertheless, the deviation between prediction and measurement is satisfactory[1]

---

[1] A better fit of the data could have been achieved if an increased absolute threshold (increased *min*-value) had been used in model **nh**. But then, the mapping to model units would also have to be changed, because an input-level according to 100 dB SPL should still lead to an output-value of 100 MU. For reasons of model universality this was not done, because a predicted absolute threshold of 0 dB SPL (at 1 kHz) is in line with data for listeners with normal-hearing in both ears.

3.4 Modulation matching

3.4.1 Method

3.4.1.1 Experimental procedure and stimuli

Modulation-matching functions presented by Moore et al. (1996) were obtained for the same subjects as reported in the previous section. The stimuli were an amplitude modulated tone at 1 kHz with a 500 ms steady state portion and 20 ms (raised-cosine) rise and fall time. The modulation was sinusoidal on a decibel scale, and modulation depth was specified as the peak-to-valley ratio in decibels. The equation specifying the stimulus waveform is:

$$F(t) = a \cdot 10^{[m(\sin(2\pi f_m + q) - 1)/40]} \sin(2\pi f_c t) \qquad (3.4)$$

Where **a** is the amplitude of the unmodulated carrier, **m** is the modulation depth (peak to valley ratio in dB), f_m is the modulation frequency, **q** is the modulator starting phase and f_c is the carrier frequency. The levels of the unmodulated carriers were chosen so that they were equally loud in both ears leading to the impression of a comfortable loudness. Essentially the same procedure was used as in the loudness-matching experiment, except that the modulation depth instead of the stimulus level was varied. The modulation rates 4, 8, 16 and 32 Hz were imposed at modulation depths of 6, 12 and 18 dB and presented as the fixed stimulus to the normal ear. For each condition, five matches were obtained by adjusting the modulation depth in the impaired ear. Then the modulation depth was fixed in the impaired ear at the mean-value so determined, and five matches were obtained by varying the modulation depth in the normal ear. As before, model predictions are compared with the data of the study by Moore et al. (1996), averaged across subjects.

3.4.1.2 Simulations

Simulations were carried out with the same model combinations as used in the first experiment (**nh-ih1** and **nh-ih2**). The level of the unmodulated carrier was chosen to correspond to 50 MU ("middle loud") within the model. This was done for all model versions leading to levels of 54 (**nh**) and 83 dB (**ih1** and **ih2**). The modulation depth (2, 4, 8, and 16 dB) of the stimulus was fixed in the model for normal hearing. The modulation depth in the models for impaired hearing was varied to determine the modulation depth at which the (ac-coupled) rms-value at the output of the modulation filter tuned to f_m matched the rms-value obtained with model **nh**. The modulation depth was varied in 0.1 dB steps within the range of 0 to 30 dB. The modulation depth whose corresponding rms-value came closest (in a least square sense) to the rms-value of the reference modulation was taken as match.

3.4.2 Results

Figures 3.3 and 3.4 show modulation-matching functions as mean data of three subjects (open symbols, replot from Moore et al., 1996) together with model predictions (filled symbols) obtained with model **ih1** (squares) and model **ih2** (circles). Figure 3.3 shows results for a modulation frequency of 4 Hz (left panel) and 8 Hz (right panel). Fig. 3.4 shows corresponding results for modulation frequencies of 16 Hz (left panel) and 32 Hz (right panel). The solid line (without symbols) indicates the expected results for two perfectly matched normal ears. For a modulation depth of 5 dB in the impaired ear, the minimum and maximum value for the modulation depth in the normal ear across subjects is indicated by the dashed lines.

In all panels, the mean data (open symbols) are shifted to higher modulation depths compared to the reference curve for two perfectly matched normal ears (solid line). This reflects the effect reported by Moore et al. (1996) that a given modulation depth in the impaired ear is matched by a higher modulation depth in the normal ear. The magnitude of this effect is roughly the same for all modulation frequencies tested. The slopes of the (best fitting) data-curves are slightly lower (4 Hz) or slightly higher (8, 16, 32 Hz) than the slope of the reference curve. Nevertheless, the authors showed that the data can well be fitted by curves with the same slope as the reference curve.

3.4 Modulation matching

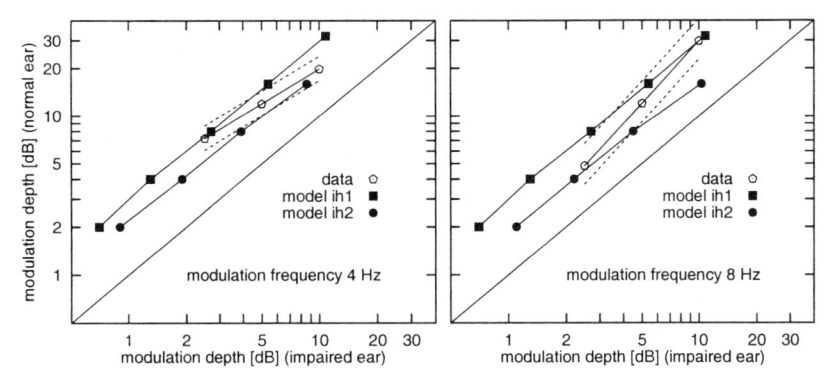

Fig. 3.3: Modulation-matching functions for modulation rates of 4 Hz (left panel) and 8 Hz (right panel). Mean measured data (open symbols) from Moore et al. (1996) for three unilateral hearing-impaired subjects and corresponding predictions (filled symbols) obtained with the combination of the models **nh-ih1** (squares) and **nh-ih2** (circles) are shown. The solid line (without symbols) indicates the expected results for two perfectly matched normal ears. For a modulation depth of 5 dB in the impaired ear, dashed lines indicate the minimum and maximum value of the modulation depth in the normal ear across subjects.

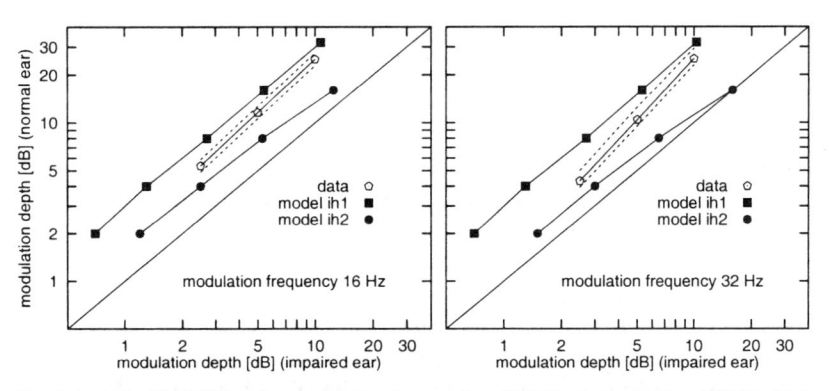

Fig. 3.4: As in Fig. 3.3, but for modulation frequencies of 16 Hz (left panel) and 32 Hz (right panel).

Similar to the empirical data, the predicted curves are also shifted to higher modulation depths and have nearly the same slope as the reference curve (1 dB/dB). The predicted modulation-matching function obtained with model **ih1** (squares) shows a stronger effect for all modulation frequencies than model **ih2**. The difference between both models is smallest for the lowest modulation frequency of 4 Hz (Fig. 3.3, left panel) and increases with increasing modulation frequency. This reflects the fact that the expansion provided in model **ih1** in independent of modulation frequency while model **ih2** provides some interaction between expansion/compression and modulation frequency. This interaction is considered in more detail in the next section. The predicted modulation-matching functions obtained with model **ih1** are more in line with the data than the predictions obtained with model **ih2**. However, modulation-matching functions for modulation frequencies > 32 Hz are necessary to clearly decide if model **ih2** fails to account for the data.

3.4.3 Analysis of model predictions

Figure 3.5 (left panel) summarizes the simulated modulation-matching functions for different modulation frequencies (replotted from Fig. 3.3 and Fig. 3.4), obtained with model **ih1** (squares) and model **ih2** (circles). Different modulation frequencies are indicated by different curve styles: „solid" (4 Hz), „long dashed" (8 Hz), „medium dashed" (16 Hz) and „short dashed" (32 Hz). Model **ih1** predicts the same modulation-masking function, independent of the modulation frequency while model **ih2** shows a strong dependence on modulation frequency (see above).

The right panel of Fig. 3.5 shows modulation transfer functions (MTFs) obtained with the model versions **nh** (triangles) **ih1** (squares) and **ih2** (circles). The ac-coupled rms-value at the modulation filter output tuned to the modulation frequency is shown as a function of the modulation frequency. Model **nh** (triangles) predicts an increasing attenuation with decreasing modulation frequency such that the shape of the MTF shows a highpass-characteristic. As a consequence, fast amplitude changes are less compressed than slow amplitude changes. The value of the MTF for a modulation frequency of about 250 Hz reaches the value 0 dB which indicates linear processing of modulations within the model **nh**. Even the adaptation loop with the shortest lowpass time constant (5 ms) acts to slow to follow, and thereby compress such rapid amplitude changes. The dashed curve indicates the MTF for the a „linear" model which transforms modulations linearly while it processes the dc-part of the stimulus in the same way as model **nh**. The MTF has a constant value of 0 dB for modulation frequencies up to about 250 Hz and decreases for higher modulation fre-

3.4 Modulation matching

quencies. The decrease is caused by the 1 kHz first-order-lowpass filter prior to the adaptation stage which is used to extract the envelope from the bandpass-filtered input waveform.

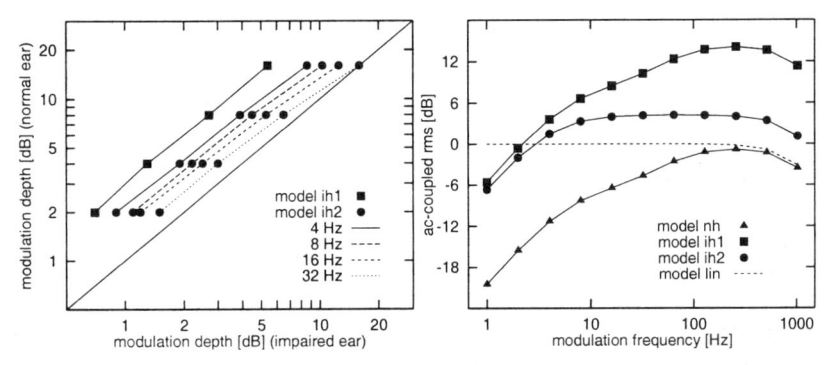

Fig. 3.5: The left panel shows predicted modulation-matching functions (replotted from Fig. 3.3 and Fig. 3.4) obtained with model **ih1** (squares) and model **ih2** (circles). Modulation frequencies of 4, 8, 16 and 32 Hz are indicated by the curve styles „solid", „long dashed", „medium dashed" and „short dashed", respectively. The right panel shows modulation transfer functions (MTFs) for the models **ih1** (squares), **ih2** (circles) and **nh** (triangles). The ordinate indicates the ac-coupled rms-value at the output of the adaptation stage. In addition, the MTF obtained with model **lin** (linear processing of modulations and the same dc-compression as in model **nh**) is indicated by the dashed curve. The MTFs were obtained with a sinusoidal amplitude modulation (m = 0.01) imposed on a dc-carrier (80 dB).

The MTF obtained with model **ih1** (squares) has the same shape as the MTF obtained with model **nh** but is shifted by 14 dB towards higher values. The difference between the MTFs obtained with model **nh** and model **ih1** is independent of the modulation frequency. Thus, the temporal properties of these models are the same because the same set of adaptation loops are used. The MTF obtained with model **ih2** (circles) also shows a highpass-characteristic. In contrast to models **nh** and **ih1**, the maximum value of the MTF is already reached for a modulation frequency of 16 Hz. The adaptation loop with the shortest time constant (50 ms) used in model **ih2** is only able to compress relatively slow amplitude changes. The difference between the MTFs obtained with models **nh** and **ih2** is nearly equal to that of model **ih1** for modulation frequencies of 1 and 2 Hz but becomes smaller for higher modulation frequencies. That is, the difference between the MTFs strongly depends on modulation frequency.

3.5 Summary and discussion

The purpose of the present study was to examine how the sensation of loudness and the sensation of modulation depth can be represented within a single model such that the perceived differences observed between normal-hearing listeners and sensorineural hearing-impaired listeners are accounted for. The basic assumption was that the reduction of peripheral compression as observed in the impaired auditory system is the main reason for these differences. A reduced overall compression was either achieved by an instantaneous expansion in model **ih1**, or by a reduction of the compressive properties of the adaptation loops in model **ih2**. These two model approaches were investigated because both realize a reduced overall compression in the processing stages associated with processing properties of the cochlea. Model **ih1** has the advantage that the reduction in compression can be adjusted by a single model parameter and that solely the compressive properties of the model are changed. However, the physiological observations seem to contradict this proposed instantaneous expansion and indicate a reduction of the compressive properties. Model **ih2** has the advantage of reducing the compressive properties in the first processing stages where a compression is realized within the model. However, for the simulation of a more severe hearing impairment, several (up to five) parameters have to be adjusted which also affect the temporal processing properties of the model.

The first experiment (Sec. 3.3) investigated the effects of the reduced overall compression on the representation of loudness within the model. The input/output functions obtained with the different model versions were interpreted as predictions of a categorical loudness scaling experiment. The values of the internal representation of the stimulus were associated with categories used in a categorical loudness-scaling experiment in a straight forward way. Model units ≤ 0 were associated with the category „inaudible" while model units ≥ 100 were associated with the category „too loud". With these assumptions, loudness-matching data for unilateral sensorineural hearing-impaired listeners by Moore et al. (1996) can be reasonably well accounted for by both models **ih1** and **ih2** if appropriate model parameter for simulating a reduced compression are chosen. That is, only the overall compression within the model has to be changed to account for this data. However, this simple approach is most likely not capable to account for other effects of loudness perception (such as loudness-summation effects or

3.5 Summary and discussion

the loudness of short stimuli) or data obtained with other measurement paradigms (e.g., such as, loudness measured in a ratio scale expressed in sone). Only data obtained with a categorical loudness-scaling experiment for steady tones or subcritical narrowband noises can be interpreted in the described way within the model.

The second simulated experiment (Sec. 3.4) examined the effects of the reduced compression on the sensation of modulation depth. The ac-coupled rms-value at the output of the optimally tuned modulation filter was used as the representation of modulation depth sensation within the model. The same parameter adjustments as in the loudness-matching experiment (Sec. 3.4) were used. Model **ih1** predicts a modulation-rate-independent increase in perceived modulation depth. This is more in line with the data by Moore et al. (1996) than the modulation-rate-dependent increase in perceived modulation depth predicted by model **ih2**. The modulation-rate-independent increase in perceived modulation depth is also in line with the assumption that the peripheral compression observed in the intact cochlea is fast-acting such that each stimulus-cycle is compressed independently and that this fast-acting compression is reduced in the impaired cochlea. Hence, the loss of peripheral compression can be modeled effectively by an instantaneous expansion. However, there appears to be a contradiction with physiological data that indicate a more linear (i.e., less compressive) input/output function of the basilar membrane.

The time constants of the adaptation loops are crucial to model temporal properties of the normal and impaired auditory system (Holube and Kollmeier, 1986, Kollmeier et al., 1997) in conditions of e.g. modulation detection and forward masking. A reduction in compression by a reduced number of adaptation loops (model **ih2**) leads to altered temporal properties. However, the temporal properties of the auditory system seem to be unaffected by sensorineural hearing loss. That is, cochlear impairment affects primarily structures *prior* to the adaptation loops which therefore represent primarily a retrocochlear compression. A more refined model of the „effective" signal processing in the auditory system should introduce an instantaneous compression, representing the peripheral compression for simulating normal hearing, combined with a certain number of adaptation loops, representing adaptation and compression effects of the retrocochlear auditory pathway. For the simulation of hearing impairment only the instantaneous compression has to be reduced.

3.6 Conclusions

- Two model versions of the modulation-filterbank model (Dau et al., 1997a) were tested for the simulation of sensorineural hearing impairment. Both lead to nearly identical loudness-matching functions but different transformation characteristics for dynamically-varying sounds (such as amplitude-modulated sinusoids). A static expansion (model **ih1**) leads to modulation-matching functions that do not depend on modulation rate. The model with altered adaptation loops (model **ih2**) predicts modulation-matching functions which are shifted towards a less "effective" loss of compression with increasing modulation rate.

- The data of Moore et al. (1996) are more consistent with the idea that loudness recruitment results from a loss of fast-acting compression (corresponding to model **ih1**). However, the variability of their data preclude an unequivocal conclusion, and further data are needed to evaluate effects of loudness recruitment on dynamic aspects of loudness. This might be also interesting for modulation rates larger than 32 Hz.

- Within the model of the „effective" auditory signal processing, sensorineural hearing loss has to be assumed prior to the adaptation stage to account for both loudness-matching as well as modulation matching data. Any change in the compressive properties of the adaptation stage is accompanied by a change of the temporal properties which would neither predict the data of the present study properly, nor the data obtained in modulation detection experiments.

Chapter 4
Modeling temporal and compressive properties of the normal and impaired auditory system[1]

Abstract

This paper presents and tests three modifications of a psychoacoustically and physiologically motivated processing model (Dau et al. 1997a). The modifications incorporate a level-dependent difference in compression between the model version for normal hearing and the model version for sensorineural impaired hearing. **Model1** realizes this difference by an instantaneous level-dependent expansion prior to the adaptation stage of the model. **Model2** and **model3** realize a level-dependent compression which is also time-dependent (time constants of 5 or 15 ms) for normal hearing and a reduced compression for impaired hearing. All model approaches account to a similar extent for the recruitment phenomenon measured with narrow-band stimuli and for forward-masking data of normal-hearing and hearing-impaired subjects using a 20 ms-2 kHz tone-signal and a 1 kHz-wide bandpass-noise masker centered at 2 kHz. A clear difference between the different model approaches occurs for the processing of temporal fluctuating stimuli. A modulation-rate-independent increase in modulation energy for

---

[1] Modified version of the paper „Modeling temporal and compressive properties of the normal and impaired auditory system", written together with Torsten Dau and Birger Kollmeier, to be submitted to J. Acoust. Soc. Am.

simulating impaired hearing is only predicted by **model1** while the other two models realize a modulation-rate-dependent increase. Hence, the predictions of **model2** and **model3** are in conflict with the results of modulation-matching and modulation-detection experiments. It is concluded that the main aspects of impaired hearing (altered loudness perception, reduced dynamic range, normal temporal properties but prolonged forward-masking effects) can effectively be modeled by incorporating a fast-acting expansion within the processing model by Dau et al. (1997a) prior to the nonlinear adaptation stage. A refined version of the processing model for normal hearing is proposed which incorporates a fast-acting compressive nonlinearity, representing the cochlear nonlinearity, followed by an instantaneous expansion and the nonlinear adaptation stage which represent aspects of the retrocochlear information processing in the auditory system.

4.1 Introduction

The processing of acoustical information in the impaired auditory system differs from the processing in the normal auditory system. In case of a sensorineural hearing loss, the alterations are at least partly due to differences in cochlear processing. Psychoacoustical results obtained with narrowband stimuli indicate a dynamically varying compressive nonlinearity acting in the healthy cochlea at the place of the characteristic frequency. Recio et al. (1998) showed that this compression is fast-acting, i.e., the compression was visible within on cycle of the center frequency at the recording site. Depending on the degree of sensorineural impairment, the cochlea reacts in a less compressive manner or even linear (Ruggero, 1992). It is assumed that the overall compression of the intact auditory system is higher than in the impaired auditory system and that the loss of compression affects each cycle of the stimulus waveform.

As was shown in Chap. 3, reduced overall compression -not necessarily fast-acting- is necessary for the modeling of recruitment observed with sensorineural hearing loss. However, in order to account for the unchanged temporal properties of the impaired auditory system as observed in modulation detection (Chap. 2) and modulation matching experiments (Chap. 3) such a reduction in overall compression has to be realized by an appropriately fast-acting processing stage.

Apart from the reduced overall compression, other alterations for modeling sensorineural hearing impairment might be necessary for a successful model of impaired hearing, such as, e.g., a general loss of sensitivity which can be accounted for by an increased variance of the internal noise (Chap. 2). Another physiological observed result which most likely influences the signal processing in the normal and impaired auditory system is the broadening of the excitation on the basilar membrane with increasing stimulus level. In notched-noise-masking experiments this effect leads to an increasing effective filter bandwidth with increasing stimulus level (Moore, 1995, Rosen and Baker, 1994). The filter bandwidth of hearing-impaired listeners are found to be generally broader than those of normal-hearing listeners (e.g. Florentine et al., 1980, Moore, 1995). However, an intricate interaction exists between the altered compressive properties in the impaired auditory system and the psychoacoustically derived parameters of temporal and spectral resolution in hearing-impaired subjects. Hence, it is still not clear which psychoacoustical parameters in impaired hearing are „primary" parameters and which parameters are „dependent"; i.e. being altered as a consequence of the primary parameters. The only way to clarify this question is to develop and validate appropriate processing models for normal hearing and to study the necessary changes in the model required to model the performance of hearing-impaired listeners.

This chapter deals with the question if and how the physiologically observed differences between a healthy and an impaired cochlea, namely the loss of the fast-acting dynamically varying compression, can be modeled on the basis of the processing model by Dau et al. (1997a+b). This model was selected since it quantitatively describes a large variety of psychoacoustical effects in normal-hearing listeners including intensity discrimination, forward-masking, temporal integration, modulation perception, spectral masking (Derleth et al., 1999) and some aspects of speech perception in normal and hearing-impaired listeners (Holube and Kollmeier, 1996).

However, this model does not assume an instantaneous compression at cochlear level but realizes an approximately logarithmic overall compression of the input level which is coupled with a time-dependent adaptation. One main question addressed in the present chapter is whether a modification of the current processing model towards a more physiologically oriented approach would still preserve the predictive capabilities for normal-hearing listeners listed above. Such a model would combine a fast-acting peripheral compression with a retrocochlear „slower" compression which still is coupled with adaptive properties in the same way as in the original model. The capabilities of two possible realizations of these modifications (referred to as **model2** and **model3**) are tested in the present study in comparison with the original model (**model1**). As the input/output function of the intact cochlea, a function presented by Moore (1997) is assumed. This function is highly compressive at medium stimulus levels while it is less compressive at low and high stimulus levels[1]. Roughly the same overall compression at medium stimulus levels has been chosen in all model versions for normal hearing. **Model2** realizes the dynamically varying compression by a generalized adaptation stage while **model3** incorporates a nonlinear compressive Gammatone-filterbank. This filterbank is in princi-

---

[1] Newer data (Recio et al. 1998) show that the cochlea acts highly compressive even at very high stimulus levels. The assumed input/output function should therefore be altered. Nevertheless, results presented throughout this paper are not critically dependent on the exact shape of the assumed input/output function.

4.1 Introduction

ple also capable to account for a broadening of filter bandwidth with increasing stimulus level while in **model1** and **model2** a linear, constant-bandwidth Gammatone-filterbank is used.

The second main question addressed in the present study is how hearing impairment might be represented within the framework of the different model modifications presented above. In each case, hearing impairment will be achieved by a reduction in compression. The goal of the study is to find out which model strategy is most successful in predicting dynamic and compressive properties in both the normal and the impaired auditory system, as observed in psychoacoustical experiments such as loudness perception, forward masking, modulation detection and modulation matching.

4.2 Models

The model approaches presented in this section are modifications of the modulation-filterbank model by Dau et al. (1997a+b). The original model consists of several preprocessing stages and an optimal detector as the decision device. The first processing stage is the linear Gammatone-filterbank model suggested by Patterson et al. (1987). After peripheral filtering, the model contains half-wave rectification, lowpass filtering at 1 kHz and an adaptation stage. The adapted signal is then processed by a modulation filterbank (Dau et al. 1997a+b). An internal noise with a constant variance is added to the output of each modulation filter to limit the resolution within the model. The signal representation at this stage of the model is referred to as its „internal representation". For the simulation of detection experiments, the optimal detector compares the internal representation of the current signal with a suprathreshold template representation of the signal (for details see App. A).

Figure 4.1 shows a schematic plot of the preprocessing stages for **model1** (left panel) **model2** (middle panel) and **model3** (right panel). The subsequent model stages following the preprocessing stages (not shown in this figure) are equal for all models and consist -as in the original model- of a modulation-filterbank, the addition of an „internal noise" and an optimal detector as decision device. The level-dependent compressive properties are realized throughout the different models in different preprocessing stages. The level-dependent compressive properties of each model approach can be adjusted to simulate normal-hearing listeners and sensorineural hearing-impaired listeners.

4.2.1 Model1

The left panel in Fig. 4.1 shows the scheme of the preprocessing stages of **model1**. This model realizes a level-dependent instantaneous expansion for simulating impaired hearing while it uses the same processing stages as the model proposed by Dau et al. (1997a) for simulating normal hearing[1]. The

---

[1] **Model1** is basically the same approach as model **ih2** presented in Chap. 3. There, an instantaneous *level-independent* expansion prior to the adaptation loops was introduced to account for the reduced compressive properties of an impaired auditory system. The *level-dependent* instantaneous expansion is realized in **model1** simply by taking the stimulus to the power of a level-dependent value.

4.2 Models

initial filtering stage is the linear Gammatone-filterbank (Patterson et al., 1987). The stimulus is further halfwave rectified and filtered by a first-order lowpass filter (1 kHz cut-off frequency). A limiter restricts each stimulus value I to be larger than or equal to a minimal value (*min*) to establish an absolute threshold within the model. Subsequently, the (envelope) signal is taken to the power of k(I). This stage was introduced to account for the differences in compression between the normal and the impaired auditory system. For simulating normal hearing, k(I) equals one, independent of the respective instantaneous value I of the stimulus. For simulating impaired hearing, the value of k(I) depends on I and assumes values larger than unity (k(I) $\geq$ 1). The function k(I) was chosen such that stimuli with low and medium levels are more expanded than stimuli at high levels. The left panel of Fig. 4.2 shows the input/output function of this initial model stage for the model version for normal-hearing subjects (filled triangles) and the model version for hearing-impaired subjects (open triangles). The input/output function for normal hearing is linear within this model. The input/output function for impaired hearing is level-dependent and matches the input range of 100 dB to an output range of approx. 155 dB where the lowest 55 dB are disregarded. The shape of this function was selected to match the inverse function to the assumed input/output function for the healthy cochlea proposed by Moore (1997). The next preprocessing stage of the model is the adaptation stage (compare Chap. 3) which combines temporal adaptation effects and compressive properties. It consists of five nonlinear adaptation loops with time constants ranging between 5 and 500 ms.

In order to align this modeling approach with the physiological results of a less compressive impaired cochlea, one has to note that the effective difference in compressive properties between the normal (compressive) and the impaired (less compressive) auditory system is well accounted for by this approach. Obviously, the input to the five adaptation loops should be interpreted as a nonlinear coded „effective" envelope information rather than the input signal's temporal envelope itself. The subsequent adaptive, time-dependent compression conducted by the five adaptation loops may then be interpreted as central processing of such an intermediate representation.

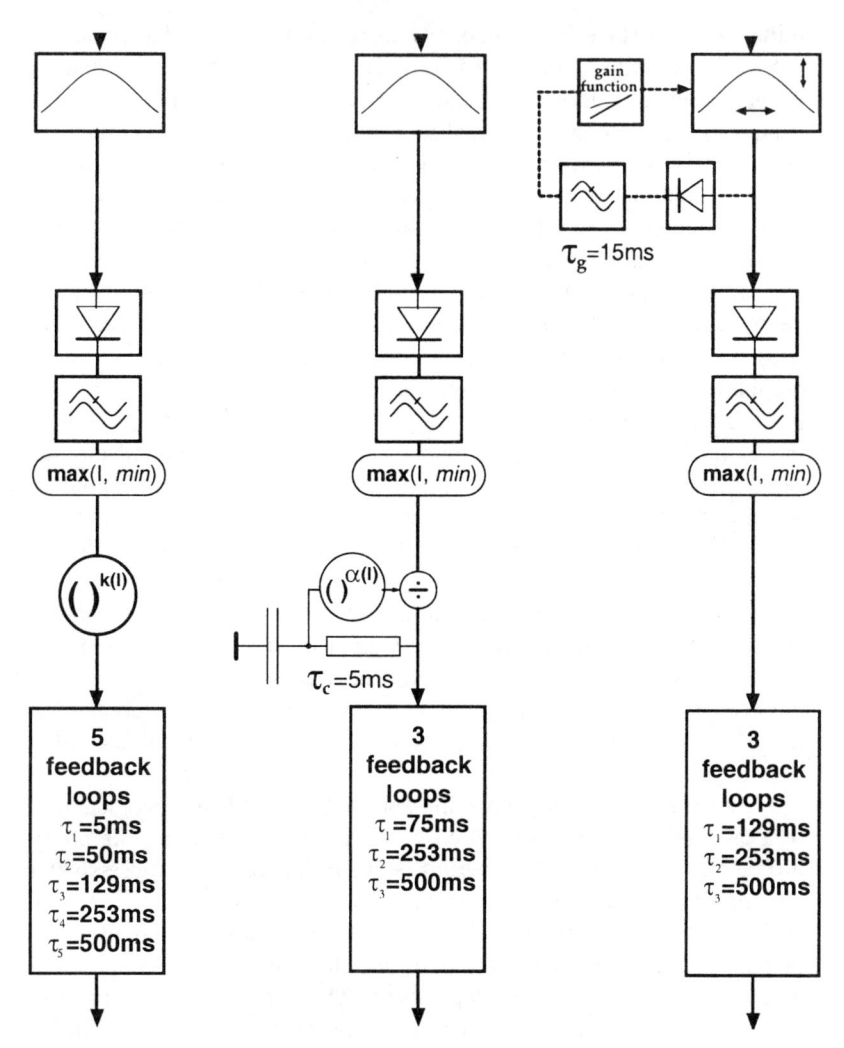

Fig. 4.1: Schematic plot of the preprocessing stages for **model1** (left panel), **model2** (middle panel) and **model3** (right panel).

4.2.2 Model2

The middle panel in Fig. 4.1 shows the scheme of the preprocessing stages of **model2**. As in **model1**, the initial stages consist of the linear Gammatone-filterbank, halfwave-rectification, lowpass-filtering and presetting a minimal value. The subsequent preprocessing stage is a modified adaptation loop which realizes the different compressive properties for the simulation of normal hearing and impaired hearing. The modified adaptation loop consists of a divider and a lowpass. The lowpass-filtered output of the adaptation loop is fed back to form the denominator of the dividing element. The divisor is the momentary charging state of the lowpass filter taken to the power of $\alpha(I)$ (I is the actual value of the stimulus-sample). The time constant is set to a value of $\tau_c = 5$ ms. For stationary stimuli, an input value I produces a value O at the output of the modified adaptation loop (derived from the stationary condition $I / O^\alpha = O$) according to the formula:

$$O = I^{\frac{1}{1+\alpha}} \tag{4.1}$$

By adjusting the value of α, this modified adaptation loop can be altered to have an expansive ($\alpha < 0$), linear ($\alpha = 0$) or compressive ($\alpha > 0$) property[1], respectively. The right panel of Fig. 4.2 shows the input/output function of this modified adaptation loop for normal hearing (filled squares) and for impaired hearing (open squares). The function $\alpha(I)$ for normal hearing is chosen such that the resulting input/output function is equivalent to the assumed input/output function for the healthy cochlea (Moore, 1997). The compression-ratio at medium stimulus levels is about 0.23 ($\alpha = 3.3$) while at extremely low and high stimulus levels it approaches a value of unity ($\alpha = 0$). For impaired hearing, $\alpha(I)$ equals zero which results in a linear input/output function of the modified adaptation loop. The subsequent preprocessing stage consists of three adaptation loops with time constants ranging between 75 and 500 ms. The number of originally five adaptation loops was reduced to the number of three adaptation loops within this model approach to keep the overall compression for the simulation of normal hearing similar to that of the original model approach. Note that the com-

---

[1] All properties of the modified adaptation loop (adaptive and compressive) are influenced by the value of α. E.g. for $\alpha = 0$, the modified adaptation loop acts linear. That is, not only stationary but also temporal-fluctuating stimuli remain unchanged by this processing stage (the time constant of this adaptation loop is meaningless in this case).

pressive properties of the modified adaptation loop at medium stimulus levels (compression-ratio: 0.23) are similar to that of two original adaptation loops (compression-ratio: 0.25).

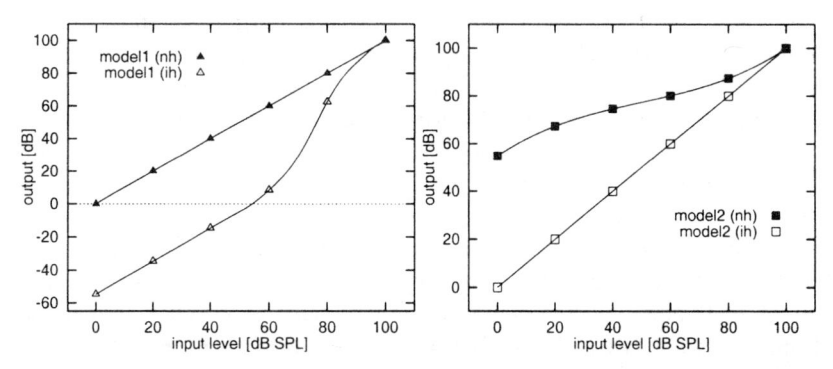

Fig. 4.2: Input-output functions of the initial model stages (prior to the adaptation stage) of **model1** (left panel) and **model2** (right panel) for simulating normal hearing (filled symbols) and for simulating impaired hearing (open symbols). The filled squares in the right panel also indicates the assumed input/output function for a healthy cochlea (Moore, 1997).

Model2 realizes a level-dependent compression for simulating normal hearing. For simulating impaired hearing, these compressive properties are reduced or even absent. This model can therefore be explained in a more straight forward way in physiological terms than **model1** if only the less compressive property of the impaired auditory system has to be accounted for. However, the physiological fact of an extremely fast-acting level-dependent compression in the healthy cochlea is not realized in this model. Instead, **model2** realizes a level-dependent compression which is time-dependent with a time constant of 5 ms. The reason for choosing such an unphysiological time constant lies in the temporal-adaptive properties of the model (cf. Sec. 4.3.3).

4.2.3 Model3

The right panel of Fig. 4.1 shows the schematic plot of the preprocessing stages of **model3**. In contrast to **model1** and **model2**, the initial stage no longer consists of the linear Gammatone-filterbank but is replaced by a level-dependent fourth-order Gammatone-filter similar to that proposed by

4.2 Models

Carney (1993). The nonlinear Gammatone-filter is designed to produce a level-dependent compression. In contrast to the linear filtering, which produces a gain of unity for a sinusoidal stimulus at the center frequency of the filter, this nonlinear Gammatone-filter produces a certain level-dependent gain. Figure 4.3 shows a schematic plot of the z-plane which illustrates how such a filter is realized. The position of the poles is a function of stimulus level and center frequency f_r. The distance between the poles and the unit circle decreases with decreasing stimulus level. The position of the pair of complex conjugate poles is indicated by the stars, plus-signs and crosses, respectively, for stimulus level of 0, 50 and 70 dB SPL. A smaller distance between the unit circle and the poles leads to smaller values of the denominator of H(z) over a wider range of frequencies. For a constant numerator value C, the predetermined gain-characteristic therefore also controls the bandwidth of the filter[1]. The left panel of Fig. 4.4 shows the input/output function for normal hearing (filled circles) and for impaired hearing (open circles). The curve for normal hearing is equivalent to the assumed input/output function of the healthy cochlea. The input/output function for impaired hearing is linear. In addition, the dotted lines indicate the amount of gain for the stimulus level of 0 (star), 50 (plus-sign) and 70 dB SPL (cross) which is produced by this filter stage for simulating normal hearing. The parameter C in the formula for the transmission characteristics H(z) is chosen such that the amount of gain becomes zero dB for a stimulus level of 100 dB SPL. The right panel of Fig. 4.4 shows the transmission characteristic of the filter stage for normal hearing at stimulus level of 0 (stars), 50 (plus-signs) and 70 dB SPL (crosses). Both panels indicate that at low stimulus levels a high gain is realized in connection with a low bandwidth and with increasing stimulus level a decreasing gain is achieved that corresponds to a higher bandwidth, respectively. The level estimate that derives the position of the poles is derived from the filter output. To do so, the filter output is halfwave-rectified and filtered by a first-order lowpass with a time constant $\tau_g = 15$ ms. This level estimate determines the gain of the filter according to the assumed input/output function (left panel of Fig. 4.4, filled circles). The effect of a sensorineural hearing loss is then modeled by removing this level-dependent compression.

---

[1] If the numerator value C is chosen to be a function of the stimulus level, any relation between filter-bandwidth and gain can be achieved. Moreover, the shape of the filter can also be changed to become asymmetric, if the position of the poles is chosen separately for each pair of complex-conjugate poles (for an example see Pflueger et al., 1997).

68   Chapter 4: Modeling temporal and compressive properties ...

position of the poles:
0 dB SPL ✻
50 dB SPL +
70 dB SPL ✕

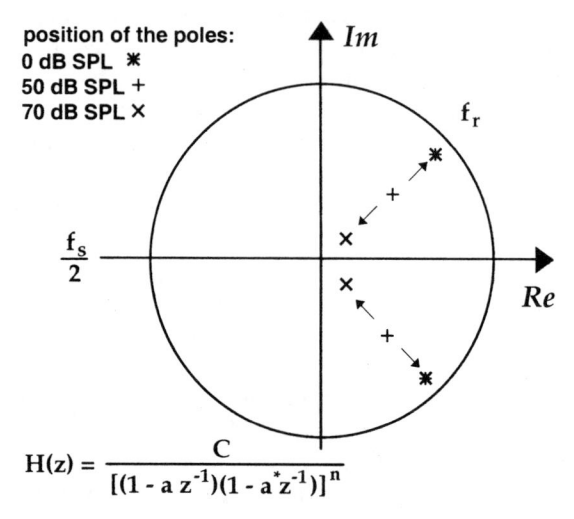

$$H(z) = \frac{C}{[(1 - a z^{-1})(1 - a^* z^{-1})]^n}$$

Fig. 4.3: Schematic plot of the z-plane for the level-dependent Gammatone-filter used in **model3**. In addition, the transmission characteristic for a filter of order n is given by H(z).

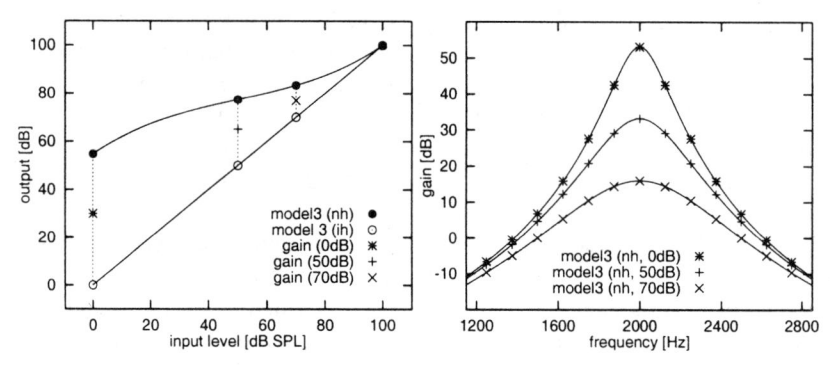

Fig. 4.4: The left panel shows input-output functions of the initial filter stage of **model3** for simulating normal hearing (filled circles) and for simulating impaired hearing (open circles). The distance between both curves represents the amount of gain and is indicated for stimulus levels of the input stimulus of 0 (star), 50 (plus-sign) and 70 dB SPL (cross). The right panel shows corresponding transmission functions for a filter centered at 2 kHz. The gain is shown as function of the stimulus frequency. Stars, plus-signs and crosses indicate the filter shape for stimulus levels of 0, 50 and 70 dB SPL, respectively.

The subsequent preprocessing stages assumed in **model3** are half-wave-rectification, lowpass-filtering and limitation to a minimal value (cf.

4.2 Models

Fig. 4.1). They are equivalent to the stages of **model1** and **model2**. The adaptation stage consists of three consecutive adaptation loops with time constants in the range of 129 to 500 ms. As in **model2**, the number of originally five adaptation loops was reduced to the number of three adaptation loops within this model approach to keep the overall compression for the simulation of normal hearing similar to that of the original model approach.

As in **model2**, the level-dependent compression realized by **model3** is time-dependent. The time constant of 15 ms is larger than in **model2** and therefore also appears to be unphysiological. Again, the reason for choosing this time constant are the temporal-adaptive properties of the model (cf. Sec. 4.3.3).

4.2.4 Simulations

Simulations were performed with the model version for normal hearing and with the corresponding model version for impaired hearing of the models **model1**, **model2** and **model3**. Simulations were carried out concerning only the output of the Gammatone-filter tuned to the respective signal frequency. For predicting thresholds, the same measurement procedure (3 AFC, 1up-2down) as in the experiments was used. The variance of the internal noise was adjusted according to the actual stimulus level to achieve a constant just noticeable level difference (intensity-JND) of 1 dB over the entire usable dynamic range for the different model versions. The details of the simulation procedure are given by Dau et al. (1996a).

For predicting the sensation of loudness, the input/output characteristic for each model version was determined. Therefore, the mean-value of the internal representation, calculated across a duration of 100 ms, is calculated for each stimulus level in the range between 0 and 100 dB SPL (step-size: 1 dB). The mean excitation is expressed in model units (MU) and associated with loudness categories of a categorical loudness-scaling experiment (Pascoe, 1978; Heller, 1985; Hohmann, 1993). Model units and corresponding verbal categories are: ≤ 0 „inaudible", 10 „very soft", 30 „soft", 50 „intermediate", 70 „loud", 90 „very loud" and ≥ 100 „too loud".

For the simulation of the sensation of modulation depth within the models, the ac-coupled rms-value of the modulation-filter output tuned to the test-modulation frequency is monitored. In modulation-matching experiments, the modulation depth whose corresponding rms-value came closest (in a least square sense) to the rms-value of the reference modulation was taken as match.

4.3 Results

The properties of the different model approaches with regard to stationary stimuli are compared in the first two sections. There, the input/output functions of the complete models and the frequency selectivity of the initial filtering stages are tested. The following two sections deal with the temporal properties of the different model approaches, i.e., simulations of a forward-masking experiment (Sec. 4.3.3) and the temporal modulation transfer functions (TMTFs) (Sec. 4.3.4).

4.3.1 Loudness functions

The left panel of Fig. 4.5 shows the input/output functions of **model1** (triangles), **model2** (squares) and **model3** (circles) for the model versions for normal hearing (filled symbols) and the model versions for impaired hearing (open symbols). A sinusoid (1 kHz) with a duration of 200 ms was used as the input signal. The output in model units (MU) increases with increasing level of the input signal for all model versions. The input/output functions of **model2** and **model3** show a level-dependent curvature for the simulation of normal hearing while **model1** shows a level-dependent curvature for simulating impaired hearing. Although the differences in the simulated curves across models are small, this different curvature reflects the two basic modeling approaches to account for the difference in level-dependent compression between normal hearing and impaired hearing. Input/output functions for simulating normal hearing approximate very well a straight line with the slope of 1 MU/dB. The input/output functions for simulating impaired hearing can qualitatively be described by two straight lines: For signal levels in the range between 0 and 55 dB SPL, the output remains constant at a value of 0 MU. For signal level higher than 55 dB SPL, the output increases with a slope of approx. 2.2 MU/dB. The similarity of the shape across models reflect the fact that all model versions for impaired hearing were tuned to simulate a sensorineural hearing loss of 55 dB and a remaining usable dynamic range of 45 dB. Since to a very rough approximation the model units in the output of the model versions can be associated with the categories of a categorical loudness-scaling procedure (cf.

4.3.2 Frequency selectivity as a function of level

The right panel of Fig. 4.5 shows the equivalent rectangular bandwidth (ERB) of a filter centered at 2 kHz as function of the stimulus level. Predictions for the three model approaches are indicated by the filled symbols. In addition, the bandwidth of level-dependent asymmetric filter proposed by Moore and Glasberg, (1987) is indicated by plus-signs and the bandwidth of level-dependent asymmetric filters proposed by Rosen and Baker (1994) is indicated by crosses and stars. In both studies, roex(p,r) shaped filter were fitted independently for the lower and the upper frequency-branch of the filter to mean data measured with the notched noise paradigm either with fixed signal-level (Ps) or with fixed masker-level (No). Thus, the abscissa values (right panel of Fig. 4.5) refer either to the signal level in dB SPL in conditions with fixed signal level (stars and filled symbols) or to the masker level in dB SPL/ERB in conditions with fixed masker level (crosses and plus-signs). The bandwidth of the derived filters differ markedly at medium and high levels[2] but approach similar values for low levels. The triangles indicate the constant ERB-bandwidth of 241 Hz of the linear Gammatone-filter which is used in **model1** and **model2** as the initial filtering stage. The circles indicate the ERB-bandwidth of the level-dependent Gammatone-filter used in **model3** for simulating normal hearing. The ERB-value of the level-dependent Gammatone-filter was determined by taking only the filter shape and not the absolute gain of the filter into account. The bandwidth increases from about 130 Hz at an input level of 0 dB SPL up to about 660 Hz at an input level of 100 dB SPL. For an input level of 50 dB SPL, the bandwidth of the linear and the level-dependent Gammatone-filter are equal.

---

[1] The standard deviation in the categorical-loudness-scaling experiment typically amounts to one category (Hohmann, 1993) which equals 10 MUs. Therefore, the differences in predicted loudness-scaling functions between **model2** or **model3** and **model1** are small in comparison to experimental errors.

[2] At least part of the difference can be attributed to the different fitting procedures used in the different studies. For fitting the filter slopes to the data, Rosen and Baker (1994) used a procedure described by the authors as PolyFit-procedure which fits all model parameters of the power-spectrum-model in once, while Moore and Glasberg (1987) used a procedure which fits only the slope of the filter.

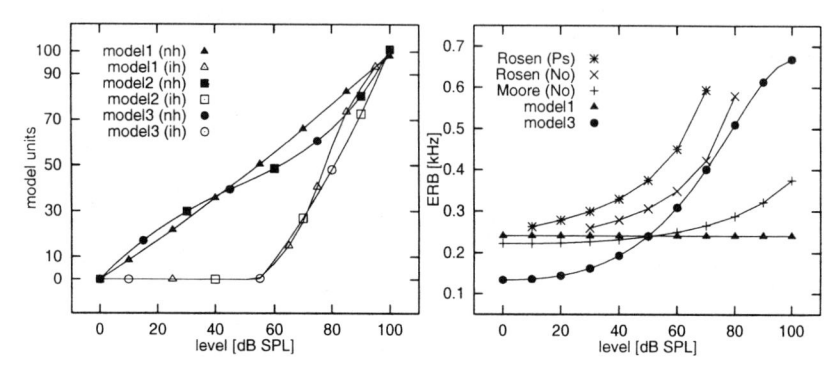

Chapter 4: Modeling temporal and compressive properties ...

Fig. 4.5: The left panel shows input-output functions for **model1** (triangles), **model2** (squares) and **model3** (circles). Filled symbols indicate calculations obtained with models for simulating normal hearing while open symbols indicate predictions obtained with models for simulating impaired hearing. The right panel shows the bandwidth of the initial filtering stage for a filter centered at 2 kHz as function of the stimulus level for **model1** and **model2** (triangles) and for **model3** (circles). Corresponding data from the literature is indicated by crosses (Rosen and Baker, 1994), stars (Rosen and Baker, 1994) and plus-signs (Moore and Glasberg, 1987).

The bandwidth of the level-dependent Gammatone-filter is too high for high stimulus levels and too low for low stimulus levels while for medium levels, the bandwidth lies in between the data reported in the previous mentioned studies. That is, the change in bandwidth with level of the proposed level-dependent Gammatone-filter is generally too high. This is at least partly due to the fact, that in case of the level-dependent Gammatone-filter both filter-branches, the upper branch and the lower branch become shallower with increasing stimulus level while in the roex-filters proposed in the study by Moore and Glasberg (1987) and by Rosen and Baker (1994) only the lower filter-branch changes. As mentioned in Sec. 4.2.3, the change in bandwidth as well as the shape of the filter could be adjusted to the data by introducing further parameters into the filterstage of the respective model. Further work is needed to clarify which changes in filter-width and filter-shape with stimulus level are appropriate within the model to account for spectral-masking data (see Chap. 5). Note, however, that the results on temporal effects presented within the remainder of this chapter are not critically dependent on the bandwidth of the initial filtering stage.

4.3.3 Forward masking

Forward-masking experiments measure the detection threshold of a signal presented temporally after a masking stimulus. The amount of forward-masking is mainly influenced by the temporal distance between masker and signal. Many stimulus parameters such as, e.g., masker level, masker and signal duration and the spectral content of the stimuli further influence the amount of forward masking. The perception model (Dau et al. 1996a) originally was designed to account for forward-masking data obtained with normal-hearing listeners. The most important component of the model to perform an appropriate prediction is the adaptation stage consisting of five subsequent nonlinear adaptation loops. A forward-masking experiment was used for adjusting the time constants of the adaptation loops to the results of a „normal" listener (Dau et al. 1996a+b). Holube and Kollmeier (1993) altered these time constants to account for impaired hearing characterized by prolonged forward masking without considering the hearing-impaired subject's reduced dynamic range.

One aim of this study was to develop modified model approaches that adequately predict forward-masking data from hearing-impaired listeners without loosing the model's predictive power for forward-masking data obtained with normal-hearing listeners. The originally used experimental condition for adjusting the time constants of the adaptation loops (**model1** for normal-hearing listeners) was also used for adjusting the time constants τ_c and τ_{1-3} of **model2** and the time constants τ_g and τ_{1-3} of **model3**. The goal was to find a combination of time constants for each model where the time constant of the initial processing stage (τ_c or τ_g) is as short as possible. This requirement results from the very fast-acting nonlinearity of the healthy cochlea which should have some correspondence in the model. The exact values of the time constants τ_{1-3} are less critical as long as they are sufficiently high to produce temporal-masking effects that last up to several hundreds of milliseconds. Since the results of the model calculations depend on the whole signal processing, it is not clear beforehand if the modified model versions (**model2** and **model3**) for normal hearing can as well account for the data as the original model version (**model1**) for normal hearing.

Figure 4.6 shows the simulated and measured thresholds for a 5 ms-2 kHz signal masked by a fixed 200-ms section of a broadband frozen-noise masker as a function of the temporal separation between masker offset and signal offset. The signal had 2.5 ms rise/fall times, shaped with a cosine-squared function while the masker was shaped by a step-function. The mean measured data of three normal-hearing subjects (taken from Dau et al., 1996) is indicated by crosses. The filled symbols connected by solid lines indicate results obtained with **model1** (triangles), **model2** (squares) and **model3** (circles), respectively. The time constants described in Sec. 4.2 were used, i.e., the initial time constants are $\tau_c = 5$ ms (**model2**) and $\tau_g = 15$ ms (**model3**). The shape of the simulated and measured curves are in good agreement. The predicted thresholds obtained with **model1** lie in general at the lower edge of the measured data and underestimate the measured thresholds for signal-masker separations ≥ 20 ms by approx. 5 dB. **Model2** and **model3** predict slightly less masking than **model1**. In addition, simulations obtained with **model3** using an initial time constant $\tau_g = 5$ ms are indicated by the filled circles connected by the dashed line. This model version predicts much too small forward masking compared to the measured data.

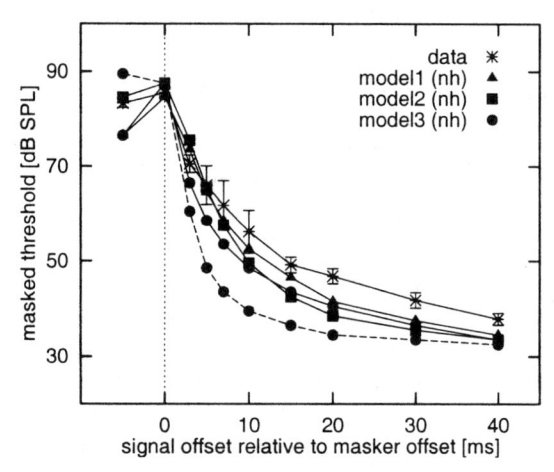

Fig. 4.6: Forward-masked thresholds of a 5 ms tone burst (2 kHz) as a function of its temporal position relative to the offset of a 200 ms broadband noise masker (20-5000 Hz, 77 dB SPL). Mean data of three subjects is indicated by stars. Model predictions obtained with model versions simulating normal hearing are indicated by triangles (**model1**), squares (**model2**) and circles (**model3**). In addition, the dashed line indicates a model prediction obtained with **model3** using a time constant $\tau_g = 5$ ms for the level-dependent initial filter stage.

This indicates that the initial time constant of **model3** can not be decreased if **model3** still should be able to account for forward-masking data measured with normal-hearing listeners. Note, however, that this large time constant

4.3 Results

is in conflict with an interpretation as basilar membrane nonlinearity, because much smaller time constants are observed physiologically at this level.

To decide whether the model approaches can account for the difference in forward-masking observed between normal-hearing listeners and hearing-impaired listeners, another forward-masking experiment reported by Glasberg et al. (1987) was simulated. The left panel of Fig. 4.7 shows their results for the experiment (same figure as in Glasberg et al., 1987; Fig. 4, lower panel). They measured the threshold of a brief (20 ms 2 kHz) signal within and following a band-limited (1.5-2.5 kHz) noise masker for two normal-hearing (filled symbols) and two sensorineural hearing-impaired subjects (open symbols) with a mean hearing-loss of 56 dB. The signal and the masker had 10 ms rise/fall times, shaped with a cosine-squared function. The time between masker-onset and signal-onset was 0, 90 and 200 ms in simultaneous-masking and 220, 240 and 280 ms in forward-masking conditions, respectively. The same masker level in dB SPL (84 dB) was used for both groups of listeners (solid lines). In addition, masked thresholds were measured at the same masker level in dB SL (49 dB SPL) for the normal-hearing listeners (dashed line).

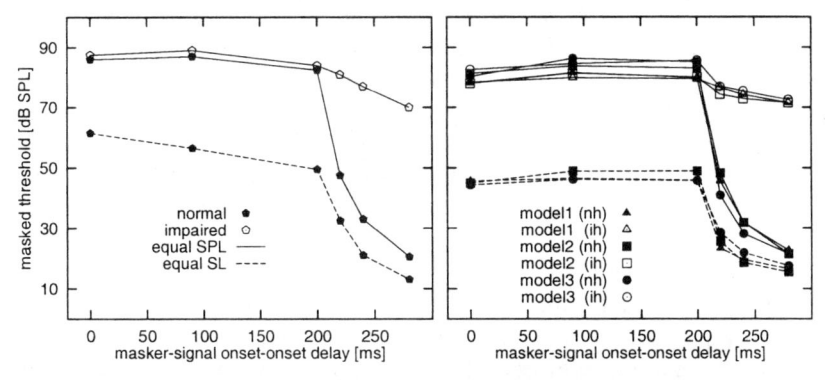

Fig. 4.7: Mean masked thresholds of a 20 ms-2 kHz tone burst as a function of its temporal position relative to a 220-ms-bandpass-noise masker. The left panel shows data for two sensorineural hearing-impaired listeners (open symbols) and for two normal-hearing listeners (filled symbols) (adopted from Glasberg et al., 1987). The solid curves indicate results obtained at the same masker level of 84 dB SPL for both groups of listeners while the dotted curves indicate results obtained at the same masker level in dB SL (49 dB SPL). The right panel shows corresponding simulations obtained with model versions simulating normal hearing (filled symbols) and model versions simulating impaired hearing (open symbols). The standard deviation of the simulated data amounts 3 dB in case of simultaneous masking (0, 90 and 200 ms) and 1 dB in case of forward-masking conditions (220, 240 and 280 ms).

The thresholds are roughly constant in conditions of simultaneous masking and decrease in forward-masking conditions with increasing time delay between signal and masker. The threshold difference between simultaneous masking and the forward-masked threshold (280 ms delay) amounts to 15 dB for the hearing-impaired listeners (open symbols) and 60 dB (35 dB) for the normal-hearing listeners at the same masker level in dB SPL (dB SL), respectively. Thus, the rate of threshold decay in forward masking is smaller in hearing-impaired listeners than in normal-hearing listeners.

The right panel of Fig. 4.7 shows corresponding model predictions for model versions simulating normal-hearing listeners (filled symbols) and for simulating impaired-hearing listeners (open symbols) obtained with **model1** (triangles), **model2** (squares) and **model3** (circles). Thresholds were simulated using nearly the same stimuli as in the measurement. In the measurement running-noise was used while in the simulation threshold estimates were obtained for 30 single frozen-noise representations. The mean value of these 30 threshold estimates was taken as predicted threshold.

All model approaches (right panel) show qualitatively the same results as the measured data (left panel). The masked thresholds remain nearly constant in simultaneous-masking conditions and decrease in forward-masking conditions with increasing signal-masker delay. The difference between masked thresholds in simultaneous and forward masking is also similar to the measured data. That is, all three model approaches can qualitatively account for the differences in forward-masking experiments between normal-hearing listeners and hearing-impaired listeners.

The main reason for the models ability to account for this difference is the reduction of the compressive properties for simulating hearing-impaired listeners. In **model1** all temporal properties remain unchanged and only the compressive properties are reduced for simulating hearing impairment. In **model2** and **model3**, the altered compressive properties also affects the temporal properties of the models in a way that the fastest acting compressive stages are less effective. That is, within these model approaches, the „effective" time constant for compression is higher for simulating hearing-impaired listeners than for simulating normal-hearing listeners. Although this effect should lead to an increased amount of predicted forward masking in the model for hearing-impaired listeners, the predictions obtained with **model2** and **model3** were very similar to those obtained with **model1**. Obviously, the reduced compression in combination with the unaltered subsequent processing stages is already sufficient to explain the prolonged recovery time in forward-masking experiments. In the study by Glasberg et al. (1987) it was already pointed out that a decrease in compression leads to a slower rate of recovery from forward masking, even if the time constant of an assumed temporal integrator subsequent to compression remains constant. As a consequence, the time constants of the model need not to be increased in order to account for the data obtained with sensorineural hearing-impaired listeners.

4.3.4 Modulation detection

One measure of the temporal properties of the auditory system is the temporal modulation transfer function (TMTF). The data for normal-hearing and hearing-impaired subjects show a similar frequency selectivity for modulations (compare Chap. 2). TMTFs were predicted reasonably well for narrowband noise carriers of the bandwidths 16 and 200 Hz, respectively, centered at 2 kHz in each condition. Fig. 4.8 shows the experimental results (replotted form Fig. 2.5) and corresponding predictions, obtained with **model1**, **model2** and **model3** for the 200-Hz bandwidth condition. The ordinate denotes modulation depth at threshold (expressed as $20\log(m)$ in dB), and the abscissa represents modulation frequency. The upper left panel of Fig. 4.8 shows the thresholds for five normal-hearing (filled pentagrams) and for four hearing-impaired subjects (open pentagrams). The remaining panels represent model predictions obtained with **model1** (upper right panel), **model2** (lower left panel) and **model3** (lower right panel). The filled symbols indicate predictions obtained with the model version for normal hearing and the open symbols indicate predictions obtained with the model version for impaired hearing.

The shape of the experimental threshold patterns (upper left panel) as well as the absolute threshold values are very similar for both groups of subjects. In this experiment, the inherent statistics of the carrier is the limiting factor for the detection of the test modulation. The frequency selectivity for modulations assumed in the models mainly results from the parameters of the modulation filterbank and the transformation characteristic of the adaptation stage For modulation frequencies above 8 Hz, thresholds increase by about 3 dB per doubling of the modulation frequency. This threshold pattern is well predicted for both normal and impaired hearing by **model1** (upper right panel). The shape of the threshold pattern obtained with **model2** (lower left panel) is similar to the data but absolute values are approx. 5 dB too low. The predicted threshold pattern for normal hearing (filled symbols) obtained with **model3** (lower right panel) is „U-shaped" and therefore not in line with the data.

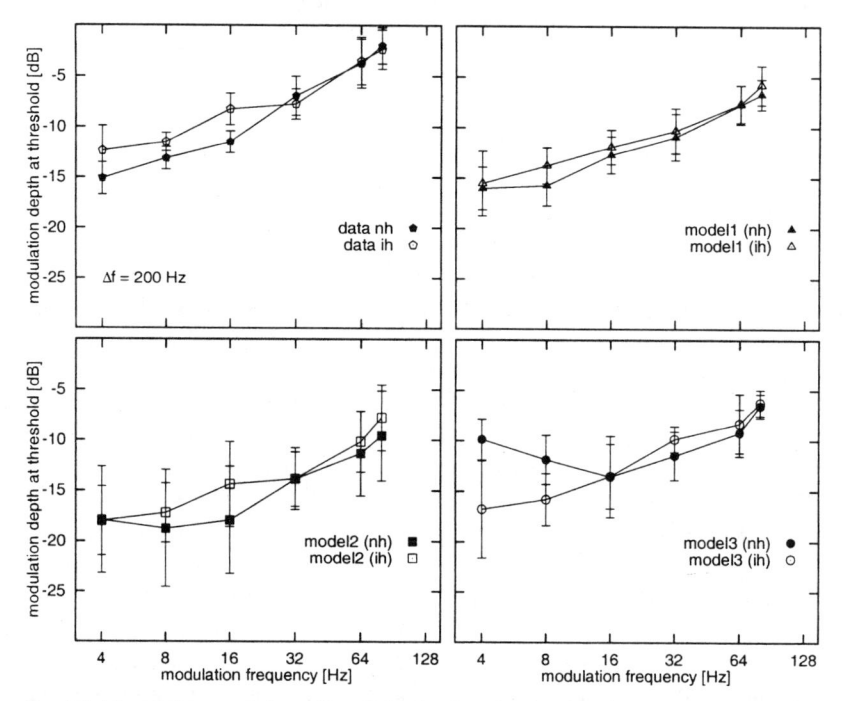

Fig. 4.8: Measured (upper left panel) and simulated modulation-detection thresholds for a bandwidth of $\Delta f = 200$ Hz, as function of modulation frequency. Predictions were obtained with **model1** (upper right panel), **model2** (lower left panel) and **model3** (lower right panel). The measured data (replot from Fig 2.5) represent mean thresholds for five normal-hearing (filled pentagrams) and for four sensorineural hearing-impaired subjects (open pentagrams). Model predictions were performed at 60 dB SPL for normal hearing (filled symbols) and 90 dB SPL for impaired hearing (open symbols) corresponding to the mean presentation level for normal-hearing and hearing-impaired listeners, respectively. Running Gaussian noise at a center frequency of 2 kHz was used. Carrier and modulation duration: 1 s.

Figure 4.9 shows the experimental results and corresponding predictions for a carrier bandwidth of 16 Hz. As for the 200-Hz condition, the experimental data are very similar for both groups of subjects. The threshold patterns show a highpass characteristic. Again, **model1** predicts the shape of the threshold pattern and the absolute values very well. The shape of the predicted threshold pattern obtained with **model2** and **model3** also shows a highpass characteristic but agree less with the data. Moreover, **model3** predicts different shapes of the threshold pattern for normal and impaired hearing which is not in line with the data.

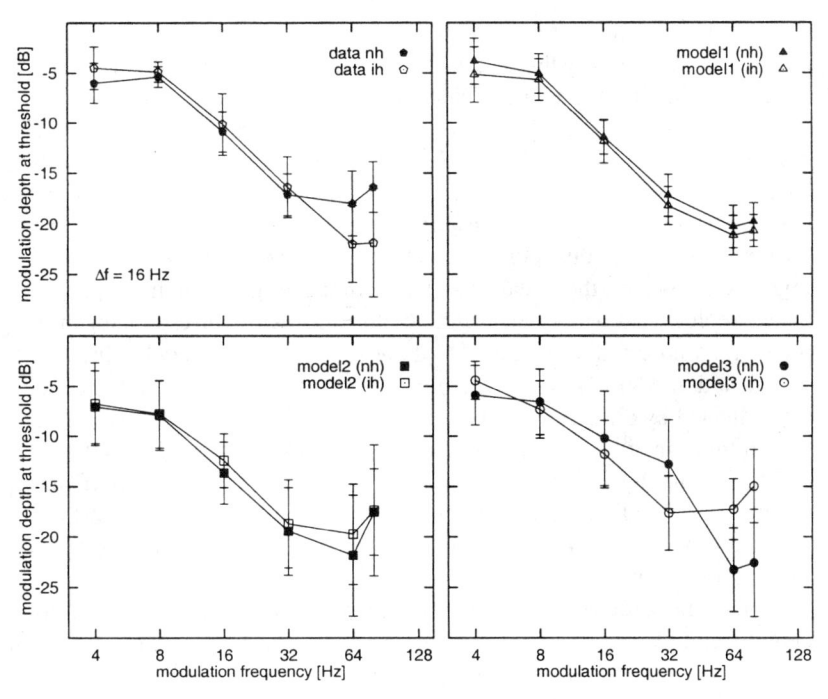

Fig. 4.9: Same as Fig 4.8 but for the Δf = 16-Hz bandwidth condition.

4.3.5 Modulation matching

In the modulation matching experiment, a direct comparison between a normal and an impaired ear for temporally-varying sounds is performed by matching the sensation of perceived modulation depth between both ears. In a study by Moore et al. (1996) the effects of loudness recruitment on steady sounds and dynamically-varying sounds were examined by loudness-matching and modulation-matching experiments with three unilateral sensorineural hearing-impaired listeners. It was shown that a given modulation depth in the impaired ear was matched by a greater modulation depth in the normal ear for modulation rates in the range from 4-32 Hz. The modulation-matching functions could be predicted reasonably well by the loudness-matching results obtained with steady tones. These results are in line with the idea that a main factor of loudness recruitment is the loss of fast-acting compressive nonlinearity that operates in the normal peripheral auditory system.

The stimuli were an amplitude modulated tone at 1 kHz with a 500 ms steady state portion and 20 ms (raised-cosine) rise and fall time. The modulation was sinusoidal on a decibel scale, and modulation depth was specified as the peak-to-valley ratio in decibels. The level of the unmodulated carrier was chosen to correspond to 50 MU ("middle loud") within the different model versions. For details about the procedure see Chap. 2.

Figure 4.10 shows the mean modulation-matching functions (upper left panel, replot from Moore et al., 1996) together with corresponding model predictions obtained with **model1** (upper right panel), **model2** (lower left panel) and **model3** (lower right panel). Different modulation frequencies are indicated by different curve styles: „solid" (4 Hz), „long dashed" (8 Hz), „medium dashed" (16 Hz) and „short dashed" (32 Hz). The solid line (without symbols) indicates the expected results for two perfectly matched normal ears. Predictions were obtained by comparing the output of the respective model version for normal hearing with the corresponding version for impaired hearing.

The mean data (upper left panel) are shifted to higher modulation depths compared to the reference curve for two perfectly matched normal ears (solid line without symbols). This reflects the effect reported by Moore et al. (1996) that a given modulation depth in the impaired ear is matched by a higher modulation depth in the normal ear. The magnitude of this effect is roughly the same for all modulation frequencies tested. The slopes of the (best fitting) data curves are slightly lower for 4 Hz and slightly higher for 8, 16 and 32 Hz than the slope of the reference curve. Nevertheless, the authors showed that the data can be fitted reasonably well by curves with the same slope as the reference curve. **Model1** predicts the same modulation-matching function, independent of the modulation frequency while **model2** and **model3** show a strong dependence on modulation frequency which is not visible within the data. Thus, only **model1** accounts for the modulation-matching data.

4.3 Results

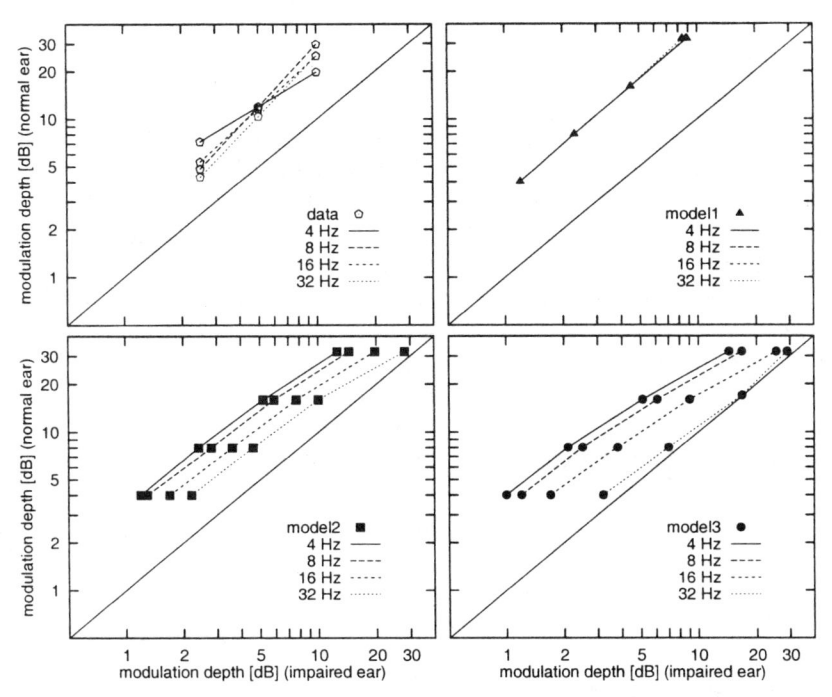

Fig. 4.10: Measured (upper left panel, replot from Moore et al., (1996)) and simulated modulation-matching functions for modulation rates of 4, 8, 16 and 32 Hz indicated by the curve styles „solid", „long dashed", „medium dashed" and „short dashed", respectively. Predictions were obtained by a combination of the model version for normal hearing and the model version for impaired hearing, using **model1** (upper right panel), **model2** (lower left panel) and **model3** (lower right panel), respectively. The solid line (without symbols) indicates the expected results for two perfectly matched normal ears.

4.3.5.1 Analysis of model predictions

To further quantify the interaction between compression and temporal processing in the models considered here, Fig. 4.11 shows modulation transfer functions (MTFs) obtained with the model versions for normal hearing (filled symbols) and impaired hearing (open symbols). The ac-coupled rms-value at the output of the adaptation stage is shown as a function of the modulation frequency. The input stimulus was a sinusoidally amplitude modulated (m = 0.01) dc-carrier at a level of 80 dB. The upper left panel of Fig. 4.11 shows the transfer characteristic obtained with **model1**. The instantaneous expansion required for modeling hearing impairment leads to a modulation-rate-*independent* increase of the output (open triangles) relative to the model for normal hearing (filled triangles). The upper and lower right panel of Fig. 4.11 shows results obtained with **model2** and **model3**, respectively. These latter two models predict a modulation-rate-*dependent* difference of the output for normal hearing and impaired hearing. In both models, the difference decreases with increasing modulation rate. For **model2** both curves converge at a modulation rate of about 100 Hz while for **model3** the curves converge at a rate of 20 Hz. The differences between the **model2** and **model3** directly reflect the different time constants of the initial processing stages which are smaller for **model2** (5 ms) than for **model3** (15 ms). Since **model1** assumes an *instantaneous* expansion for the simulation of hearing impairment, the output curves for normal and impaired hearing are simply shifted to one another.

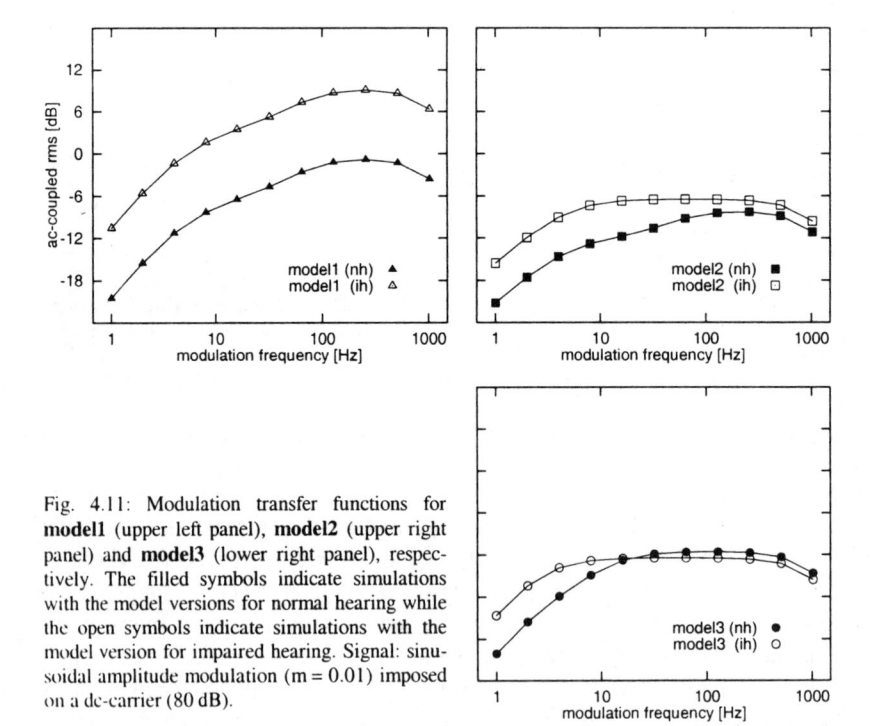

Fig. 4.11: Modulation transfer functions for **model1** (upper left panel), **model2** (upper right panel) and **model3** (lower right panel), respectively. The filled symbols indicate simulations with the model versions for normal hearing while the open symbols indicate simulations with the model version for impaired hearing. Signal: sinusoidal amplitude modulation (m = 0.01) imposed on a dc-carrier (80 dB).

4.4 Discussion

The purpose of this chapter was to examine the interaction between compressive and temporal properties of a model for normal hearing and the necessary changes in order to account for hearing impairment. The underlying approach is to split up the compressive properties of the model in two independent stages: The first stage should mimic the compressive properties of the peripheral (cochlear) auditory system while the second stage should reflect some retrochochlear compression which affects mainly stationary stimuli parts. Three models were presented: **Model1** combines an instantaneous compression and a consecutive expansion for normal hearing with five nonlinear adaptation loops. In the model version for normal hearing, the compression and expansion effectively cancel each other so that a linear filterbank remains. In the model version for impaired hearing, the instantaneous compression is missing. That is, only the (level-dependent) expansion is combined with the same five nonlinear adaptation loops as in the model for normal hearing. **Model2** combines a modified adaptation loop which realizes a level-dependent compression with a set of three adaptation loops for normal hearing. In the model version for impaired hearing, the modified adaptation loop acts less compressive. **Model3** combines a nonlinear (compressive) Gammatone-filterbank as the initial processing stage with a set of three adaptation loops for normal hearing. In the model version for impaired hearing, the nonlinear filterbank acts less compressive.

All model versions can account for the recruitment effect (Sec. 4.3.1) and forward-masking data of normal-hearing and hearing-impaired subjects (Sec. 4.3.3). They differ in frequency selectivity (Sec. 4.3.2), their sensitivity for modulations (Sec. 4.3.4) and in their physiological interpretation.

Model3 can be excluded, because it was shown that this model approach can only account for forward-masking data of normal-hearing listeners if an unphysiological long time constant (15 ms) is chosen for the initial filtering stage. As a result of this long time constant, this model is not able to predict a modulation-rate-independent difference in the outputs of the model versions for normal and impaired hearing (Sec. 4.3.5.1) which is needed to predict modulation-matching experiments (Sec. 4.3.5). A predicted difference in modulation energy which is less dependent on modulation-rate can be achieved by a smaller time constant of the initial filtering-stage. However this would also lead to altered temporal-adaptive properties of the model such that forward-masking experiments could not be accounted for any more. Similar problems can be expected if the adaptation stage would be combined with nonlinear transmission-line models (e.g. van Hengel, 1996, Meddis, 1999) as the initial model stage. One way to overcome this problem would be to replace the adaptation stage of the model by a processing stage which acts less sensible to already compressed input stimuli than the set of adaptation loops. Nevertheless, such an alternative adaptation stage would need to account for the basic signal processing properties of the set of adaptation loops, namely a nearly logarithmic compression of stationary stimulus parts and a less compressive processing of temporally fluctuating stimulus parts.

Model2 can be excluded for similar reasons as **model3**. This model approach can also only account for forward-masking data of normal-hearing listeners if an unphysiological long time constant (5 ms) is chosen for the initial adaptation loop. As a result of this long time constant, this model is not able to predict modulation-matching experiments.

It remains **model1** which accounts to a similar extent as **model3** (and **model2**) for the recruitment effect and for forward-masking data of normal-hearing and hearing-impaired subjects. Moreover, this model accounts also for a modulation-rate-independent increase in modulation energy which allows to predict modulation-matching and modulation-detection data obtained with normal-hearing and hearing-impaired subjects. These results are achieved by using the original processing model for simulating normal hearing combined with an instantaneous expansion prior to the adaptation stage for simulating impaired hearing. That is, it seems to be sufficient to realize a reduced overall compression for simulating impaired hearing by a fast-acting processing stage to account for some basic perception differences between normal-hearing and hearing-impaired listeners. The question is if and how such an „effective" expansion is realized in the auditory pathway and why it only affects the signal processing in the impaired auditory system.

The normal human auditory system is able to perceive physical sound events within a dynamic range of approximately 100 dB at least for frequencies in the range of 1-4 kHz. Stimuli within this dynamic range cause basilar membrane movements in the healthy cochlea of mammals within a markedly smaller dynamic range (about 50 dB). Therefore, the compressive nonlinearity in the healthy cochlea transforms the wide dynamic input range into a smaller output range. Moreover, single nerve fibers are not capable to encode this already reduced dynamic range by the rate of nerve impulses. At least a number of nerve fibers with markedly different absolute

4.4 Discussion

thresholds are needed to further transmit the stimulus information (Yates, 1990, Patuzzi, 1992). Other encoding mechanisms (e.g. simultaneous firing of nerve-fibers) may also play a role. Nevertheless, on a perceptive level of the auditory system, the whole dynamic range of the input signal is available in some way. It can be assumed that the information concerning the dynamic range of the stimulus is reconstructed from the signals of several nerve fibers. In the end, by combining (e.g. summing with different weights) informations of several nerve-fibers, the effects of the peripheral compression on the dynamic range of the input stimulus obviously is compensated for on higher processing stages (Zeng et al., 1997). It is still unclear how and at which stage of the auditory pathway this may happen. Since the exact realization of the information processing in the auditory pathway is beyond the scope of a model of the „effective signal processing", it may be sufficient to model this simply by an expansion.

Fig. 4.12 shows a scheme of such a proposed model for the effective signal processing in normal-hearing and hearing-impaired subjects which includes such an effective expansion: The initial processing stage consists of a nonlinear transmission line model (alternatively, a nonlinear filterbank which compresses the stimulus nearly instantaneously may be used). This stage splits the signal into several channels which are processed in parallel. This stage represents the frequency-place transformation, the critical band concept, the level-dependent compression and the level-dependent change in excitation produced by the basilar membrane movement. The stimulus is further half-wave rectified to mimic the transformation from basilar membrane movement into receptor potentials of the inner hair cells. Up to this stage, the stimulus waveform represents the change in sound pressure either in the air or in the fluids of the inner ear. The halfwave-rectified signal is further expanded by a function which effectively acts as the inverse to the compression introduced by the initial processing stage. Within this stage, the realization of the stimulus represents somehow combined information (activation) of several nerve fibers. The following processing stages are the same as in the processing model by Dau et al. (1997a), e.g. lowpass filtering, nonlinear adaptation stage, modulation filterbank and decision device. For modeling impaired hearing, the initial nonlinear processing stage is modified such that it acts less compressive or even linear, representing the loss of peripheral compression observed in the impaired cochlea.

Model1 can bee seen as a first order approach to such a model. In the model version for impaired hearing, the compression, representing the compressive properties of the cochlea, is absent and effectively all stimuli are therefore expanded by the subsequent stage. In the model version for normal hearing, the combination of cochlear compression and subsequent expansion by the inverse function effectively does not change the stimuli. Therefore, both stages are excluded from the model for normal hearing which is then equal to the original processing model proposed by Dau et al. (1997a).

basilar - membrane filtering		reduced compression and reduced spectral resolution
halfwave rectification		
lowpass filtering		
expansion		
absolute threshold		increased absolute threshold
adaptation		
modulation filterbank		
		increased internal noise
decision device	optimal detector	

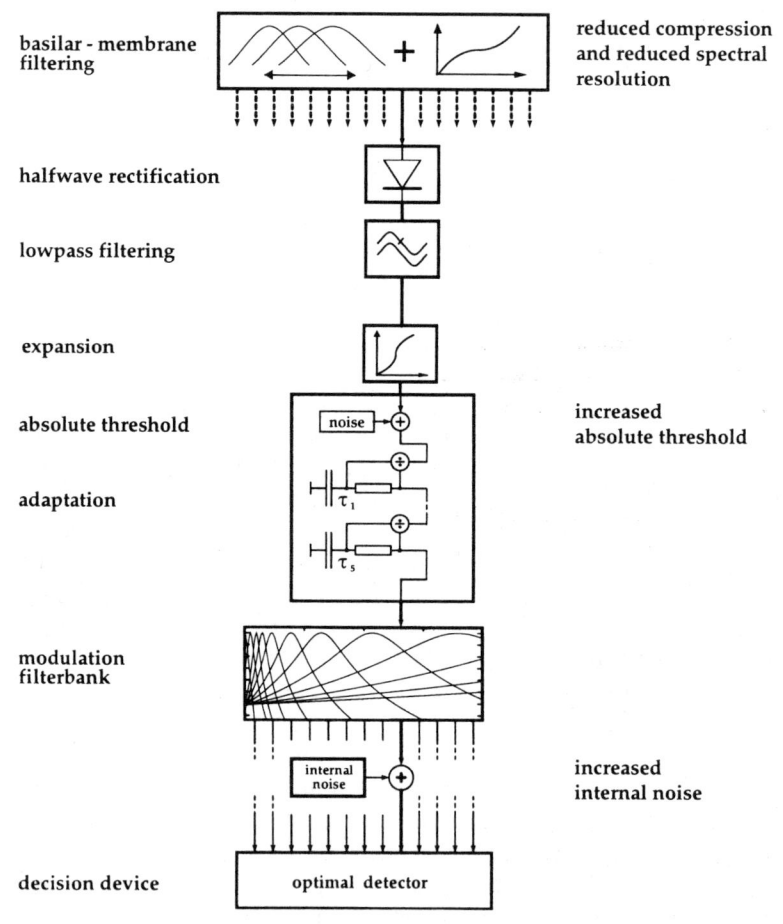

Fig. 4.12: Model structure for modeling the „effective" signal processing in the normal auditory system (middle column) and explanation (left column). The right column indicates the processing stages and the necessary changes for modeling sensorineural impaired hearing.

4.5 Conclusions

- Three model approaches were compared that include different kinds of level-dependent compression for simulating normal-hearing and hearing-impaired listeners. All model approaches can account at least qualitatively for the recruitment effect and the differences in forward masking between normal-hearing and hearing-impaired listeners. The models differ with respect to their spectral resolution as a function of level and the interaction between temporal resolution and compressive nonlinearities.

- Only **model1** which realizes an instantaneous expansion for simulating hearing-impaired listeners predicts a sufficiently high modulation-rate-independent difference in modulation energy between the model for normal hearing and the model for impaired hearing. Therefore, only this model approach is able to also account for modulation-matching data.

- The use of a linear Gammatone-filterbank in the processing model for normal hearing (Dau et al. 1997a) can be interpreted as a first order approximation of the combined processing properties of a peripheral compressive nonlinearity (cochlea) and a subsequent reconstruction of the original stimulus dynamic (i.e. an expansion) on higher stages in the auditory pathway.

Chapter 5
On the role of envelope modulation processing for spectral masking effects[1]

Abstract

This chapter examines the role of temporal cues such as beatings and intrinsic envelope fluctuations in spectral masking. Model predictions on the basis of the processing model by Dau et al. (1997a, b) are compared with averaged experimentally derived masking patterns obtained by Moore et al. (1998). In these experiments, tones and narrowband noises have been used as the signal and the masker, so that all four signal-masker combinations are considered here. In addition, model predictions in conditions of notched-noise masking are compared with own experimental data. The model uses Gammatone-filters as peripheral filtering stage and analyses envelope fluctuations by a modulation filterbank at the output of each peripheral channel. For low and medium masker levels, the model accounts very well for the masking patterns in all signal-masker conditions, as well as for the notched-noise conditions. In contrast, additionally performed predictions on the basis of an energy-detector model mainly account for the notched-noise data, but not for the shape of the masking patterns. For the high masker level, the simulations suggest the use of an asymmetric filter, with steeper high-frequency slope as assumed in the linear model. In addition, peripheral nonlinearities become apparent at this masker level, while they do not seem to play a role at lower levels.

---

[1] Modified version of the paper „On the role of envelope modulation processing on spectral masking effects", written together with Torsten Dau, submitted to J. Acoust. Soc. Am.

5.1 Introduction

Spectral masking effects in the human auditory system have been investigated over a long period of time by many researches using different psychoacoustic measurement paradigms. The threshold patterns obtained in the typical experimental masking condition were often primarily considered as *spectral* effects exclusively explained in terms of the spectral shape of the auditory filters, or alternatively, in terms of the excitation pattern on the basilar membrane produced by the particular masking stimulus. However, it also has been argued that, depending on the experimental paradigm, also other than solely spectral effects may contribute to the spectral masking data. This chapter deals with the question to what extent the processing of temporal envelope fluctuations (modulations) in the auditory system contributes to the data normally associated with spectral masking. A psychoacoustical processing model (Dau et al., 1997) which performs both spectral and temporal analysis for detecting a signal is used for describing experimental results in two different „classical" masking paradigms: One measures the detectability of a narrowband signal in the presence of a narrowband masker, as a function of the spectral separation between both. The resulting threshold function for the signal is called *masking pattern*, also called masked audiogram. The „notched-noise" paradigm measures the detectability of a tone in the presence of two masking noise bands, placed on either side of the signal spectrum as a function of the spectral notch-width in the masker. The basic assumption in either measurement paradigm is, that the masked threshold reflects the amount of excitation evoked by the masker at the signal frequency. Although modeling efforts have been undertaken, up to now, no model is available which can account for the whole set of data observed with these two measurement paradigms.

Masking patterns for narrowband signals, either sinusoids or narrowband noises, have been derived in a number of studies (e.g., Wegel and Lane, 1924; Fletcher and Munson, 1937; Egan and Hake, 1950; Zwicker, 1956; Ehmer, 1959, Greenwood, 1971; Buus, 1985; Mott and Feth, 1986; Zwicker and Fastl, 1990; Moore et al., 1998). All studies showed systematic differences in the shape of the patterns obtained with sinusoidal and narrowband noise maskers, respectively: Generally, masking patterns obtained with narrowband-noise maskers are much more regular than those obtained with sinusoidal maskers. Masking patterns obtained with sinusoidal mask-

5.1 Introduction

ers often have irregularities, i.e., dips and peaks above the masker frequency. Masking for signal frequencies below the masker frequency is smaller for sinusoidal maskers than for the corresponding narrowband-noise maskers. This is also the case for signal frequencies just above the masker frequency while more masking is found for signal frequencies well above the masker frequency. Besides the frequency separation between the signal and the masker and the physical characteristics of the stimuli, masking patterns are influenced by a variety of more complex factors. For example, thresholds may be influenced by the occurrence of combination tones produced by the nonlinearity on the basilar membrane, by aural harmonics, by beats (envelope fluctuations), and by the ability of the auditory system to „listen in the dips" of the masker envelope (e.g., Buus, 1985; Moore and Glasberg, 1987; van der Heijden and Kohlrausch, 1995; Nelson and Schroder, 1996).

Recently, Moore et al. (1998) investigated the relative importance of the factors influencing the masking patterns for sinusoidal and narrowband-noise maskers. As the signal they used both sinusoids and narrowband noises, so that all four masker-signal combinations were considered. The same subjects participated in the different experimental conditions and the same experimental procedure was used in each condition. Several experiments were undertaken by the authors to explore the role of the cues which (might) influence the shape of the masking patterns. Despite a large inter-individual variability in the data, Moore et al. (1998) found some general properties of the masking patterns. The shape of the patterns for signal frequencies above and below the masker frequency were determined mainly by the characteristics of the masker rather than the characteristics of the signal. They found that tone maskers produce less masking than noise maskers for frequency separations of 100-250 Hz between the signal and the masker. They also showed that beats between the signal and the masker presumably contributed to the effect of less masking with tone maskers compared to noise maskers both below and above the masker frequency. When the signal frequency was equal to the masker frequency, thresholds were similar for all masker-signal combinations except for the tone masker and the noise signal, which gave much lower thresholds. The authors stated that the lower threshold in the latter case can be attributed to the availability of a within-channel cue of a fluctuation in level. Another main effect influencing the shape of the masking patterns on the high-frequency side at high masker levels has been shown to be caused by the occurrence of combination products.

A major problem of the study by Moore et al. (1998) is that it still remains unclear to what extent the different effects contribute to the shape of the measured masking patterns. No modeling effort has been undertaken to clearly separate between these effects and to quantify the role of statistical properties of the stimuli on the shape of the masking patterns. This is attempted in the present chapter. Model predictions are presented on the basis of the processing model suggested by Dau et al. (1997a+b). This model has originally been developed for describing temporal masking effects such as forward masking and detection of test tones in broadband noise (Dau et al. 1996a+b). Later on, a more general version of the model has been applied to modulation-detection and modulation-masking conditions (Dau et al., 1997a+b). As an essential part of its preprocessing, the latter model contains a modulation filterbank at the output of each peripheral filter. This stage allows the processing of envelope fluctuations with rates up to about 1 kHz and enables the model to use beats (in this frequency region) as a detection cue. As peripheral processing stage, the model contains the linear Gammatone filterbank (Patterson et al., 1987). Thus, predictions obtained with such a model may offer the possibility to separate between effects caused by beating and/or modulation detection and modulation masking and those effects associated with peripheral nonlinearity such as combination products. The results obtained with the modulation filterbank model will be compared with results obtained with a „classical" energy detector model which has often been used in the literature, (e.g., Fletcher, 1940, Patterson and Moore, 1986) for describing masked thresholds.

Temporal effects probably play only a minor role in the second masking paradigm to gather information about the frequency selectivity of the auditory system, the notched-noise paradigm. Most studies used broad noise bands as the masker (e.g. Patterson, 1987) to preclude other than spectral cues produced in the investigated frequency region from detection (e.g. spread of excitation, the occurrence of distortion products). Because of the random inherent envelope fluctuations of the broad noise bands, it can be expected that also cues produced by the occurrence of beats (e.g. between the signal and the edge of the masking noise bands) are masked. This is tested quantitatively in the present study: Broad noise bands as well as narrow noise bands are used as the masker to reveal the role of envelope modulations (e.g. beats) in notched-noise masking conditions. The same models as in the previous experiment are used for predicting the data.

5.2 Experiment I: Masking patterns for sinusoidal and noise maskers and sinusoidal and noise signals

5.2.1 Method

5.2.1.1 Procedure and Stimuli

Model predictions were obtained by simulating the measurement procedure described in Moore et al. (1998). An adaptive three-interval forced-choice (3IFC) procedure was used. The signal was gated on synchronously with the masker in one of the intervals, chosen randomly. After three successive correct responses, the signal level was decreased, while after a single incorrect response it was increased (1up-3down algorithm). This procedure tracks the signal level corresponding to 79.4% correct. The initial step size was 8 dB. After two reversals the step size was halved until the minimal step size of 1 dB was reached. Eight further reversals were obtained and the threshold was estimated as the mean of the signal levels at this eight reversals. The mean value of several of those threshold estimates was taken to obtain a „predicted threshold". In conditions with a tone masker, 2 to 3 threshold estimates were taken resulting in a standard deviation of the predicted threshold of ≤ 2 dB, which was the criterion to accept the predicted threshold as valid. In conditions with a noise masker, 3 to 5 threshold estimates were needed to satisfy the same criterion.

All stimuli were digitally generated at a sampling frequency of 44.1 kHz with the aid of the signal processing software SI (developed at the University Göttingen). Either a tone or a 80-Hz wide running Gaussian noise was used as the signal and the masker. Both were gated synchronously (10 ms raised-cosine rise/fall times) and had a steady-state duration of 200 ms. Simulated masking patterns for masker levels of 45, 65 and 85 dB are shown in this study. The masker center frequency was 1 kHz. Signal frequencies of 0.25, 0.5, 0.75, 0.9, 1.0, 1.1, 1.25, 1.5, 2.0, 3.0 and 4.0 kHz were used. All four masker-signal combinations were tested (**nn:** noise-signal, noise-masker; **nt:** noise-signal, tone-masker; **tn:** tone-signal, noise-masker; **tt:** tone-signal, tone-masker). In the **tt**-condition, a 90 degree phase-shift between signal and masker was chosen while random phase-shifts were used in all other conditions. In case of the noise stimuli, a randomly generated noise sample out of a cyclic narrowband noise of 1.5 seconds duration was used for each stimulus presentation. The spectral shaping of the noise was performed in the frequency domain by setting all bins outside the desired frequency range to zero.

5.2.1.2 Simulations

Simulations were performed with the modulation-filterbank model by Dau et al. (1997a+b). The model consists of several preprocessing stages and an optimal detector as the decision device. The first processing stage is the linear Gammatone-filterbank model suggested by Patterson et al. (1987) that was originally developed to account for spectral masking data obtained with the notched-noise paradigm (Patterson, 1976). The filter shapes are symmetric on a linear frequency scale and the filter bandwidths are independent of stimulus level. The spacing of the filters is chosen in a way that the filter slopes overlap at their -3 dB points. The parallel processing of stimulus information in different peripheral channels enables the model to integrate information across frequency assuming independent observations at the stage of the detector. This allows to account for effects of off-frequency listening. In simulations of the present study all peripheral channels ranging from half an octave below to one octave above the signal frequency are considered. After peripheral filtering the model contains half-wave rectification, lowpass filtering at 1 kHz and an adaptation stage (for details see Chap. 2). The adapted stimulus at the output of each channel is then processed by the (linear) modulation filterbank (Dau et al. 1997a+b). To limit the resolution within the model, internal noise with a constant variance is added to the output of each modulation filter. As a modification of the original model, the center frequency of the highest available modulation filter was restricted to half the center frequency of the peripheral channel but never exceeded 1 kHz. Finally, the subsequent optimal detector compares the internal representation of the current signal with a suprathreshold template representation of the signal. In the present chapter, the template was calculated as the mean representation of eight suprathreshold signal-representations. The suprathreshold signal level for deriving the template was adjusted in an adaptive procedure (see App. C for details).

Simulations were also run with a modified model referred to as „modulation-lowpass model" which only processes the output of the lowest modulation filter in each peripheral channel. This modulation filter is a second-order-lowpass filter with a cut-off frequency of 2.5 Hz. Thus, this

5.2 Experiment I: Masking patterns for sinusoidal and noise maskers ... 95

processing strategy comes close to an energy-detector model (cf. Dau et al., 1996a+b).

5.2.2 Results

Figures 5.1 to 5.4 show the masking patterns for the four signal-masker configurations tested. The masker level was 85 (circles), 65 (diamonds) and 45 dB SPL (squares), respectively. The ordinate represents masking as the difference between masked threshold and absolute threshold at the specific signal frequency. The abscissa indicates signal frequency on a logarithmic scale. The left panels of Figs. 5.1-5.4 show mean data for three normal-hearing subjects (solid curves with open symbols, replot from Moore et al., 1998) together with model predictions (dashed curves with filled symbols) obtained with the modulation-filterbank model. The right panels of Figs. 5.1-5.4 show predictions obtained with the modulation-lowpass model (solid curves with filled symbols). For direct comparison, the simulations obtained with the filterbank model are replotted from the left panel (dashed curves with filled symbols).

5.2.2.1 Masking patterns obtained with a tone masker

Figure 5.1 shows masking patterns for the tone-signal tone-masker (**tt**) condition. The measured data (open symbols) show a sharp tuning around the masker frequency. Masking decreases with increasing signal-masker frequency separation except for the signal frequency of 2 kHz at a masker level of 85 dB, where the pattern shows a peak. For masker levels of 45 and 65 dB, the simulations (filled symbols) agree very well with the experimental data for all signal frequencies tested. Threshold differences are smaller than 5 dB. In particular, the model accounts for the sharp tuning around the masker frequency. Discrepancies between data and simulations only occur at the highest masker level. On the low-frequency side, masking is overestimated by 9 and 13 dB at the signal frequencies 0.5 and 0.75 kHz, respectively. On the high-frequency side of the pattern, masking is underestimated by 15 and 8 dB at the signal frequencies 2 and 3 kHz, respectively.

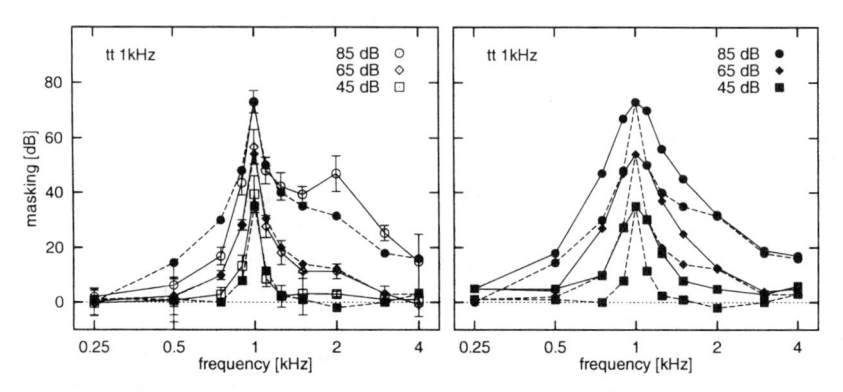

Fig. 5.1: Masking patterns for the tone-signal tone-masker (**tt**) condition, for masker levels of 85 (circles), 65 (diamonds) and 45 dB SPL (squares). The masker frequency was 1 kHz and the signal frequencies were 0.25, 0.5, 0.75, 0.9, 1.0, 1.1, 1.25, 1.5, 2.0, 3.0 and 4.0 kHz, respectively. The left panel shows mean experimental data (solid curve with open symbols) for three normal-hearing listeners (replot from Moore et al., 1998) and predicted data (dashed curve with filled symbols) obtained with the modulation-filterbank model. The right panel shows model predictions obtained with the modulation-lowpass model (solid curves with filled symbols). For direct comparison, the simulations from the left panel are replotted (dashed curves with filled symbols).

The right panel of Fig. 5.1 shows the predicted masking on the basis of the modulation-lowpass model (solid curves). For equal signal and masker frequency, the amount of masking is the same in both models. This shows that the processing of higher-frequency envelope fluctuations is not advantageous in this specific condition. No beating occurs between signal and masker, only the increase in intensity can be used as a detection cue[1]. The masking patterns obtained with the modulation-lowpass model show a much broader tuning at all three masker levels than the patterns obtained with the modulation-filterbank model. Obviously, the processing of envelope fluctuations (beats) is most advantageous for signal-masker frequency separations up to 500 Hz. The masking predicted by the two models differs by as much as 20, 16 and 10 dB for signal-masker frequency separations of 100, 250 and 500 Hz, respectively. These differences are independent of the

---

[1] A similar experimental condition (Dau et al., 1996a, 1997a) has been chosen for adjusting the variance of the internal noise in the two model versions. The calibration of the model is based on the 1 dB-criterion in intensity-discrimination tasks at medium stimulus levels. In the experimental conditions of the present study, the signal-to-masker ratio at threshold is about -7, -6 and -5 dB for the masker levels of 85, 65 and 45 dB SPL, respectively. This corresponds to predicted just noticeable differences (JNDs) of 0.8, 1.0 and 1.2 dB according to the formula: JND = $10 \log_{10} (1 + \Delta I/I)$.

5.2 Experiment I: Masking patterns for sinusoidal and noise maskers ... 97

masker level. For separations ≥ 1 kHz both models lead to the same estimates.

Figure 5.2 shows results for the noise-signal tone-masker (**nt**) condition. The experimental masking patterns are very similar in shape and absolute values as the patterns from the **tt**-condition from Fig. 5.1. The only exception is the amount of masking for equal signal and masker (center) frequency which is much lower for the noise signal than for the tone signal. Again, at masker levels of 45 and 65 dB, the simulations agree very well with the measured data for all signal frequencies tested. In particular, the model accounts for the reduction in masking (up to 18 dB) when signal and masker have the same frequency. As in the **tt**-condition, major differences between the model predictions and the data occur at the masker level of 85 dB. Again, masking is overestimated by as much as 13 dB at a signal frequency of 0.75 kHz on the low-frequency side of the pattern, while it is underestimated by 14 and 10 dB at signal frequencies of 2 and 3 kHz on the high-frequency side.

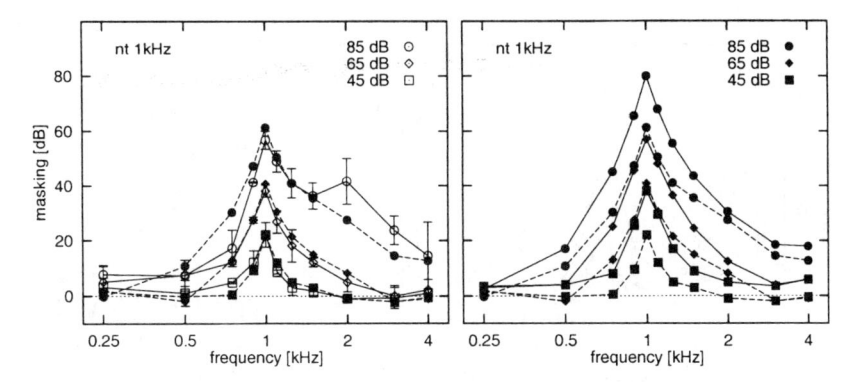

Fig. 5.2: As in Fig. 5.1, but for a noise-signal and a tone-masker (**nt**-condition).

The right panel of Fig. 5.2 shows the corresponding patterns obtained with the modulation-lowpass model. Masking is considerably increased at almost all signal frequencies compared to the results obtained with the modulation analysis. For signal-masker separations ≤ 500 Hz, threshold differences between both model predictions amount to 10-20 dB. These differences can mainly be attributed to beating between the tone-masker and the edge-frequencies of the noise-signal that are exploited in the modulation-filterbank model but can not be used as detection cues in the modulation-lowpass model. The predictions given in Fig. 5.2 are comparable to those in the **tt**-condition. Note, however, that in the on-frequency situation of the **nt**-condition, the modulation-filterbank model correctly predicts the empirical threshold which is much lower than in the corresponding **tt**-condition. Within the model this is caused by the random intrinsic envelope fluctuations of the noise signal which give an additional detection cue in the signal interval. This cue is not accessible in the modulation-lowpass model. These intrinsic fluctuations are also the reason for the small threshold differences of 3-5 dB between the models for spectral separations > 1 kHz.

5.2.2.2 Masking patterns obtained with a noise masker

Figure 5.3 shows results for the tone-signal noise-masker (**tn**) condition. The tuning of the experimental masking patterns (open symbols) is broader than in the **tt**-condition (Fig. 5.1). For example, for a signal-masker frequency separation of 100 Hz, masking is about 10 dB higher in the **tn**- than in the **tt**-condition.

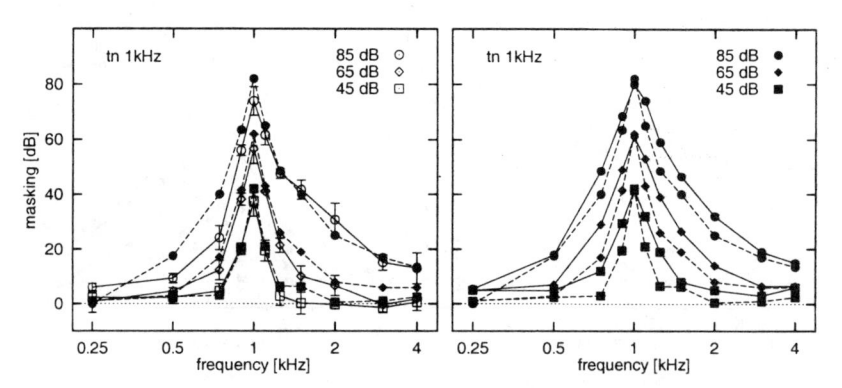

Fig. 5.3: As in Fig. 5.1 and 5.2, but for a tone-signal and a noise-masker (**tn**-condition).

This is the case for all masker levels. On the other hand, for the masker level of 85 dB and for signal frequencies ≥ 2 kHz, masking is *smaller* than in the **tt**-condition (see Moore et al., 1998). If one compares the on-frequency threshold for the **tn**- and the **nt**-condition, „asymmetry in masking" can be observed: The tone-signal is masked by the noise-masker (**tn**-condition) much more effective than the noise-signal by the tone-masker (**nt**-condition). This is consistent with recent results by Hall et al. (1997)

5.2 Experiment I: Masking patterns for sinusoidal and noise maskers ...

who investigated masking for a set of bandwidth conditions for the signal and the masker having the same center frequency. For the masker levels of 45 and 65 dB, the simulations (filled boxes in the left panel) agree very well with the measured data except for the signal frequency of 1.5 kHz where the model overestimates masking by about 7 and 10 dB, respectively. For the high masker level of 85 dB, the simulations agree very well with the measured data for signal frequencies larger than 1 kHz while too much masking is predicted on the low-frequency side of the pattern (similarly to the results in the previous two conditions) as well as in the on-frequency condition.

The masking patterns obtained with the modulation-lowpass model (solid curves in the right panel of Fig. 5.3) show a broader tuning at all masker levels than the masking patterns obtained with the modulation-filterbank model (dotted curves). The two curves differ by up to 12 dB for signal-masker frequency separations ≤ 250 Hz. For higher signal-masker frequency separations the difference in predicted masking decreases with increasing signal frequency. In the on-frequency situation, both models predict nearly the same masking. The processing of higher-frequency envelope fluctuations is not advantageous within the filterbank model in this condition, mainly the dc-component of the modulation spectrum at the output of the peripheral channels is used for detection.

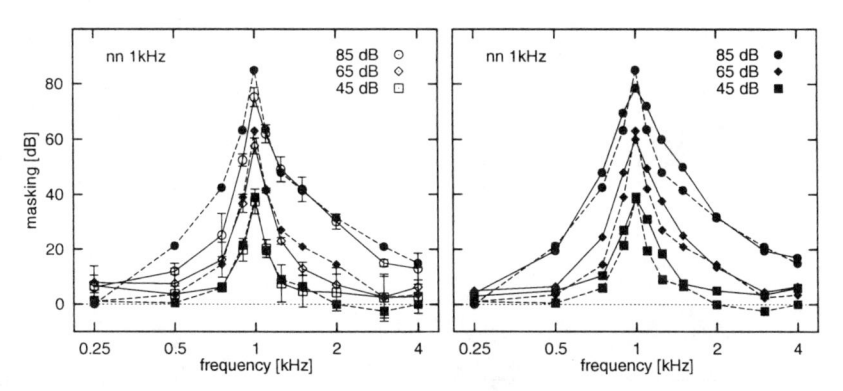

Fig. 5.4: As in the previous figures, but for a noise-signal and a noise-masker (**nn**-condition).

Finally, Fig. 5.4 shows the results for the noise-signal noise-masker (**nn**) condition. The experimental masking patterns (open symbols) are very similar in shape and absolute values to those obtained in the **tn**-condition

(Fig. 5.3). For the lowest masker level, the simulations (filled boxes in the left panel) agree very well with the experimental data at all signal frequencies. For the medium masker level, the agreement is well except for the signal frequencies 1.5 and 2 kHz, where the model overestimates masking by about 8 dB, similarly as in the **tn**-condition for 1.5 kHz signal frequency. For the high masker level, simulations agree very well at the high-frequency side of the pattern, while masking is overestimated at the low-frequency side of the pattern as well as in the on-frequency condition.

The right panel of Fig. 5.4 shows the differences between the modulation-filterbank model and the modulation-lowpass model. These differences are similar to those found in the **tn**-condition (Fig. 5.3). In the on-frequency situation of the **nn**-condition, both models predict the same masking for the lowest masker level, while for the medium and high masker level the modulation-filterbank model predicts 4.5 and 6.5 dB more masking than the modulation-lowpass model, respectively. These differences result from the specific properties of the adaptation loops in response to stimulus onsets in combination with the model's high sensitivity for envelope fluctuations[1].

---

[1] If two narrowband noise stimuli (such as the signal and the masker in the **nn**-condition) of the same duration, center frequency, bandwidth and with similar envelopes are added, very large amplitudes can occur at the output of the adaptation loops. In these specific situations, the large onset leads to an internal representation of the stimulus which is less correlated with the stored (mean) template representation as normally obtained at this signal level, resulting in a higher threshold at the end of the experimental run. This „artifact" of the model is much weaker or absent for lower masker levels and is not found for larger masker bandwidths (Verhey et al., submitted 1999).

5.3 Experiment II: Notched-noise masking

5.3.1 Method

5.3.1.1 Procedure and Subjects

Masked thresholds were measured and simulated using an adaptive three-interval forced-choice (3IFC) procedure. The masker was presented in three consecutive intervals, separated by silent intervals of 500 ms. In one randomly chosen interval the test signal was gated on synchronously with the masker. The subject's task was to specify the interval containing the test signal. The threshold was adjusted by a 1up-2down algorithm (Levitt,1971), which converges at a signal level corresponding to a probability of being correct of 70.7%. A run was started with the signal level about 20 dB above the expected threshold value. The step size was 8 dB at the start of a run and was divided by 2 after every two reversals of the signal level until the step size reached a minimum of 1 dB, at which time it was fixed. Using the 1-dB step size, ten reversals were obtained and the median value of the signal levels at these ten reversals was used as the threshold value. The subjects received immediate feedback during the measurements. For each subject the experimental run was repeated three times for each signal configuration. The mean of the threshold estimates was taken as threshold value. Three trained male normal-hearing subjects (audiometric thresholds within 10 dB HL) participated at the experiment. They were between 30 and 33 years old.

5.3.1.2 Apparatus and Stimuli

All acoustic stimuli were digitally generated at a sampling frequency of 44.1 kHz. The on-board 16-bit D/A converters of a Silicon Graphics INDY work station were used to transform digital to analog stimuli which were attenuated with the aid of a programmable amplifier and presented monaurally via headphones (Sennheiser HDA200) in a single-walled soundproof booth (IAC). Stimulus generation and presentation of trials were controlled by a computer using the signal-processing software package SI. The subjects responded via a PC keyboard.

The masker used in this experiment was a Gaussian noise consisting of two noisebands, separated by a spectral notch linearly centered at 1 kHz. The notch width was 0, 100, 300 and 500 Hz, respectively. The bandwidth of each noise band on either side was 640, 320, 160, 80, 40 and 20 Hz. Each noise interval had a steady-state duration of 100 ms and was randomly cut out of a 3 s long noise waveform to perform a running-noise experiment. Spectral shaping of the noise bands was achieved by setting all bins outside the derived frequency range in the 3 s-noise waveform to zero. The signal was a 1 kHz sinusoid with the same duration as the masker. The starting phase of the signal was generated with a 90-degree shift relative to the 1 kHz frequency-component of the masker. Each stimulus was gated with 10 ms $\cos^2$-shaped ramps. The presentation level was 68.5 dB SPL.

5.3.1.3 Simulations

Simulations were carried out with the same model versions as used in the first experiment (modulation-filterbank model and modulation-lowpass model). They were run in the same frequency region around 1 kHz as in the masking pattern experiments from Sect. 5.2. The same procedure for deriving the threshold as in the real experiment (1up-2down) was used. In addition, simulations were compared with results obtained with an energy-detector model. This model approach simply calculates the masker-energy which falls within a Gammatone-filter tuned to the signal frequency.

5.3.2 Results

The upper left panel of Fig. 5.5 shows mean experimental data for three subjects (open symbols). The other panels of Fig. 5.5 show model predictions (filled symbols) obtained with the modulation-filterbank model (upper right panel) and those obtained with the modulation-lowpass model (lower left panel) and with the energy-detector model (lower right panel). Masked thresholds are plotted as a function of the notch-width. Parameter is the masker bandwidth Δf on either side of the notch, indicated by the different symbols.

In all panels, thresholds decrease continuously with increasing notch-width. The rate of decrease is about 5 dB/100 Hz in the experimental data (upper left panel), independent of the noise bandwidth. This is consistent with corresponding notched-noise masking data by Patterson (1976) obtained with masker bandwidths larger than 640 Hz. The modulation-

5.3 Experiment II: Notched-noise masking

filterbank model (upper right panel) accounts well for the shape of the experimental threshold-curves, which may not be surprising since the shape of the Gammatone-filters has been determined by Patterson (1986) in corresponding notched-noise masking experiments. The predicted thresholds on the basis of the modulation-lowpass model (lower left panel) and on basis of the energy-detector model (lower right panel) also decrease with increasing notch-width at the same rate as in the data. Thus, as expected, the overall energy at the output of the Gammatone-filter centered at the signal frequency accounts for the general shape of the threshold-curve. Inherent envelope fluctuations of the noise masker do not seem to play a dominant role for the notch-width effect.

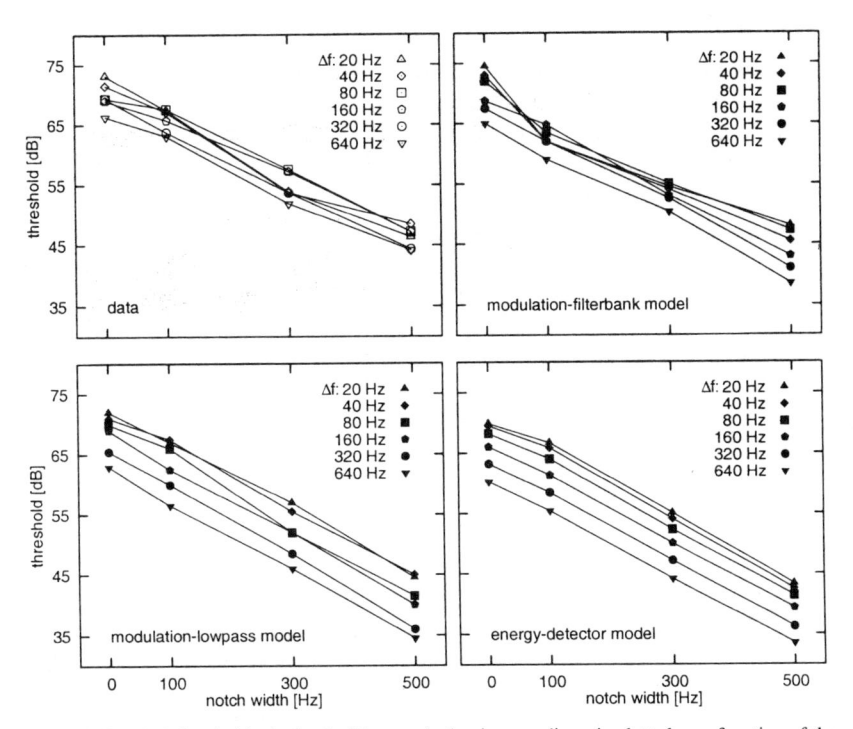

Fig. 5.5: Masked threshold, obtained with a notched-noise paradigm, is plotted as a function of the notch-width. Mean experimental data for three subjects (upper left panel) and model predictions obtained with the modulation-filterbank model (upper right panel), obtained with the modulation-lowpass model (lower left panel) and obtained with the energy-detector model (lower right panel) are shown. The bandwidth Δf of the masking noise-bands are indicated by the different symbols. The mean standard deviation of the experimental data is 3 dB.

However, distinct differences between the different model predictions can be observed for a fixed notch-width. Since the overall level (SPL) of the masker was held constant in this experiment, the energy-detector model predicts a 3 dB-decrease in threshold per doubling of the noise bandwidth as long as the frequency range of the noise exceeds the critical bandwidth (about 160 Hz) of the Gammatone-filter tuned to the signal frequency of 1 kHz. For noise bandwidths ≤ 80 Hz, the decrease is less than 3 dB. Basically the same predictions are derived by the modulation-lowpass model. Contrary to these predictions, the experimental data shows only a 1.5 dB-decrease in threshold per doubling of the noise bandwidth independent of the notch-width. The modulation-filterbank model accounts for the observed 1.5 dB-decrease for notch-widths of 100 and 300 Hz. In these conditions, signal detection within the modulation-filterbank model is strongly influenced by the inherent masker fluctuations: Thresholds are lower than those obtained with the modulation-lowpass model, whereas in the other conditions (notch-widths of 0 and 500 Hz), both models lead to nearly the same predictions. It should also be noted that the differences across model predictions in this paradigm are much smaller than for the paradigms described before (Sec. 5.2). This shows that temporal cues are less dominant in the notched-noise conditions than in the masking conditions with a single noise band.

5.4 Discussion

The purpose of the present study was to examine the role of temporal cues in experiments commonly associated with spectral masking. Two „classical" masking conditions, notched-noise masking and masking patterns were considered. Data obtained with these experimental paradigms are primarily considered to represent *spectral* effects explained in terms of the spectral shape of auditory filters and on the basis of the excitation pattern produced on the basilar membrane (spread of excitation). However, as also discussed in previous studies, asymmetries exist in the data which can not be explained solely on the basis of spectral properties, e.g. the sharply tuned peaks in the masking patterns in the **tt**-condition and the „asymmetry of masking" between the **tn**- and the **nt**-condition. The model predictions shown here based on temporal envelope detection cues in bandpass-filtered stimuli, can resolve these contradictions without loosing the models properties of predicting notched-noise data in which temporal cues obviously play a minor role. This shows that a spectro-temporal analysis is needed rather than a purely spectral analysis to account for the large range of effects observed with both measurement paradigms.

For low and medium masker level, the correspondence between predicted and experimental masking patterns is very large. For the highest masker level, the model predicts too much masking on the low-frequency side of the pattern in *all* signal-masker configurations. These results obtained with the linear Gammatone-filter whose high-frequency slope apparently is too shallow are consistent with results by Moore (1995) and suggest a level-dependent high-frequency slope of the filter. Such a filter would also reduce the predicted masking in the on-frequency conditions obtained with the noise maskers (**tn**- and **nn**-condition). The effects on the high-frequency side of the masking patterns are less clear. Experimental data are different for noise- and tone-masker which can not be accounted for by the model on the basis of temporal effects. It remains unclear if and to what extent peripheral nonlinearities of the auditory system contribute to these differences. One possible explanation for the increased masking found with tone-masker may be the occurrence of aural harmonics.

The individual frequency selectivity of normal-hearing listeners may differ markedly. This variability across subjects is also visible in the data by Moore et al. (1998). Also the strength and salience of combination products may be different for the individual listeners because of the different fine-structure of the absolute threshold (Long, 1984, Mauermann et al., 1999). Individual masking patterns are therefore probably influenced more strongly by peripheral nonlinearities and the individual shape of the auditory filters than the averaged masking patterns. However, it can be assumed that the temporal information of envelope fluctuations is available to every listener in the same way. Since the Gammatone-filters account only for the mean frequency selectivity of normal-hearing listeners, the simulations in the present study were compared with the averaged data of Moore et al. (1998). The results show that for low and medium masker levels, detection cues produced by peripheral nonlinearities are not needed to account for these averaged masking patterns and that temporal cues are clearly dominant.

The model does not include any spectral weighting to account for the mean absolute threshold for normal-hearing listeners. The model predicts an absolute threshold that is nearly independent of the signal frequency. Since masking denotes the difference between the masked threshold and the absolute threshold, the predictions are not directly influenced by the absolute threshold value. A difference in absolute threshold values between the signal- and the masker-frequency would also not influence the predictions, since the detection advantage by the temporal cues does not depend on masker level within the model (see right panel of Fig. 5.1). Hence, a (linear) frequency weighting according to the transfer characteristic of the middle-ear would not affect the predicted masking in the model.

5.5 Conclusions

- The modulation-filterbank model suggested by Dau et al. (1997a+b) uses a combined spectro-temporal signal analysis which appears necessary and meaningful to explain the different data sets obtained with both experimental paradigms, masking patterns and notched-noise masking.

- For low and medium masker levels, the model accounts very well for masking patterns obtained with narrowband signals and maskers as well as for notched-noise masking data.

- Within the model, peripheral filtering by the Gammatone-filterbank in combination with the processing of higher-frequency envelope fluctuations (such as intrinsic fluctuations and beatings between signal and masker) can lead to sharply tuned masking patterns (such as, e.g., in tone-on-tone masking conditions), while it essentially leads to similar predictions as an energy-detector model in conditions where envelope fluctuations play a minor role (such as, e.g., in noise-on-noise and notched-noise masking conditions).

- For high masker levels, an asymmetric peripheral filter with a steeper high-frequency slope as that of the Gammatone-filters would have to be assumed to account for the data. In addition, peripheral nonlinearities influence the shape of the masking patterns. Such nonlinearities are not included in the model.

Chapter 6
Summary and concluding remarks

In the first part of this thesis (Chap. 2-4) a model was developed for predicting the performance of hearing-impaired human listeners in psychoacoustical tasks. The processing model by Dau et al. (1997a+b) for normal hearing was modified such that the compressive properties of the model are reduced for simulating impaired hearing. This was motivated by the physiological observation of a level-dependent compressive nonlinearity in the healthy cochlea which acts less compressive in the impaired cochlea (Ruggero, 1992). It is assumed that the difference in compressive properties of the normal and the impaired auditory system is the primary factor responsible for the deteriorated performance of hearing-impaired listeners in psychoacoustical tasks. Moreover, it is assumed that the recruitment effect, i.e. the distorted relation between loudness perception and stimulus intensity, is a direct consequence of the reduced compression in the impaired auditory system.

To clarify the influence of altered compressive properties on the temporal acuity of the system, modulation detection and modulation masking were investigated (Chap. 2). The use of narrowband-noise or sinusoidal carrier allowed to obtain frequency-specific information of the temporal acuity which is essential in case of a frequency-specific hearing impairment. The stimuli were chosen such that the spectral content of the stimuli could not be used as a detection cue. The presentation level was chosen to produce an equal-loud sensation of an comfortable loudness in each individual. This minimizes the direct influence of the altered loudness perception because it guarantees that all parts of the stimuli are audible and that no part of the stimuli produces an uncomfortable loudness which both may lead to a deteriorated performance. In modulation-detection and modulation-masking tasks with stochastic carriers, basically the same performance of both auditory systems (normal and impaired) was observed. In these conditions, the external statistics of the stimuli mainly determines the detection-threshold of a test-modulation. Both test- and masker-modulation are processed either by the normal or by the altered compressive nonlinearity which results in the same detection threshold. In modulation-detection tasks with sinusoidal carriers (deterministic), the internal statistics of the auditory system is the limiting factor. Increased modulation-detection thresholds were observed for two out of four hearing-impaired listeners. A certain amount of increased modulation-detection threshold was always accompanied by a similar increased intensity-discrimination threshold. This indicated that the reduced ability to detect intensity differences and not a reduced temporal acuity is the reason for the increased modulation-detection thresholds. It was concluded that the temporal processing properties of the normal and the impaired auditory system are nearly equal. The model for normal hearing and the model for impaired hearing therefore should have the same temporal processing properties. Moreover, for simulating some hearing impaired subjects, a generally reduced sensitivity has to be assumed within the model which is realized by an increased variance of the „internal" noise.

In order to model hearing impairment, two approaches were presented which realize a reduced compression compared to the model for normal hearing (Chap. 3), independent of the stimulus level. The influence of the reduced compressive properties which were either realized by an instantaneous expansion or a reduced number of adaptation loops on the processing of static and temporally varying stimuli were investigated. Experiments on matching of perceived loudness and matching of perceived modulation depth between a normal and an impaired auditory system (Moore et al., 1996) were considered. The difference in compressive properties of both model versions (normal and impaired) accounts for the steeper loudness-matching function between the normal and the impaired auditory system. The association of mean excitation produced by the stimulus within the model with loudness categories allows to predict the recruitment effect. The prediction of an increased impression of modulation depth, observed in the impaired auditory system, can only be accounted for by the model which realizes a reduced compression by a structure which is extremely fast-acting, namely the model which incorporates an instantaneous expansion for modeling impaired hearing.

The realization of a reduced compression for impaired hearing by an expansion seemed to be contradicted by the physiological result of an extremely fast-acting level-dependent compressive nonlinearity in the healthy cochlea (Recio et al., 1998) and the loss of it in the impaired cochlea which led to the quest for different model approaches (Chap. 4). A model for normal hearing which incorporates a nonlinear Gammatone-filter as initial filtering stage was presented. This nonlinear Gammatone-filter realizes a level-dependent gain by a level-dependent pole-shift and therefore also the bandwidth of the filter changes with level. The proposed input/output function for a healthy cochlea (Moore, 1996) was used as the gain-function. The simulation of impaired hearing was achieved within this model by a reduced amount of gain which leads also to broadened filters. More work would have to be done to introduce a realistic functional connection between the predicted gain and the change in filter-width and to change the shape of the filter to become asymmetric (cf. Pflüger et al., 1997). Such a modeling approach is a good candidate for a filterbank which incorporates many effects one would like to account for in the initial stage of a psychoacoustical processing model. Unfortunately, it is not possible to combine such a fast-acting nonlinear Gammatone-filter, or any other processing stage which incorporates a fast-acting compression (e.g. a nonlinear transmission-line model), with the adaptation stage of the model by Dau et al. (1997a+b). The temporal adaptive properties of the model are only in line with the data obtained in a forward-masking experiment if the nonlinear Gammatone-filter acts so slow, that the prediction of other effects, namely the prediction of modulation-detection and modulation-matching data, becomes impossible. One way to solve this problem is to replace the adaptation stage of the model by another stage which on the one hand reacts less sensible on precompressed stimuli but on the other hand keeps the basic processing properties of the adaptation loops, namely the asymmetric processing of stationary and fluctuating stimuli-parts. This asymmetric processing enables it to account with the same detection criterion for both, intensity discrimination and modulation detection (Chap. 2).

As an alternative, the model approach which realizes an instantaneous expansion (independent of level) for simulating impaired hearing was modified such that a level-dependent expansion was realized (basically the inverse of the gain-function used for modeling the normally working cochlea). It was shown that this model also accounts for the recruitment effect and modulation matching. The use of the identical adaptation stage in the model for normal and impaired hearing results in identical temporal processing properties of both models which is consistent with results from modulation detection and modulation masking (Chap. 2). Moreover, this model accounts for forward-masking data obtained with normal-hearing listeners and hearing-impaired listeners. Without changing any time constant within the model, prolonged recovery times in forward-masking tasks and a reduced dynamic range between simultaneous- and forward-masked thresholds are predicted by the model for impaired hearing (Chap. 4).

Summarizing these results, the processing model by Dau et al. (1997a+b), modified by an instantaneous expansion and, if necessary, an increased variance of the internal noise for simulating impaired hearing, accounts for a couple of fundamental psychoacoustically observed processing capabilities of the normal and the impaired auditory system. The processing model for normal hearing which does not incorporate a peripheral compression may be interpreted as a first order approach of the combined processing properties of such a peripheral compression with the same subsequent expansion as in the model for impaired hearing. A more refined processing model for normal hearing should therefore incorporate a nonlinear transmission line model or a (fast-acting) nonlinear filterbank (similar to the one presented in Chap. 4) as the initial processing stage in combination with a subsequent stage with expansive properties prior to the nonlinear adaptation stage of the model. The use of a nonlinear transmission line model as the initial processing stage would introduce some effects which are so far not accounted for by the model, e.g., the level-dependent frequency selectivity and the occurrence of distortion products observed in the normal auditory system.

To account for the level-dependent frequency selectivity in the processing model used so far, the width of the initial filter would have to be adjusted separately for normal and impaired hearing according to the stimulus level or a related quantity. To quantify which relation between filter width and stimulus level is appropriate within the model for normal hearing, experiments on masking patterns and notched-noise masking were investigated (Chap. 5). Masking patterns were predicted and compared with mean experimental data (Moore et al. 1998) for all four signal-masker combinations which can be made up of sinusoidal and narrowband signals and sinusoidal and narrowband maskers. It was shown that the processing model (modulation-filterbank model) which processes also higher-frequency modulations accounts astonishing well for masking patterns obtained with all signal-masker combinations at low and medium masker levels. An alternative model approach which processes mainly the energy of the stimulus fails to account for the sharp tuning in the tone-on-tone condition and can also not account for the asymmetry of masking between the tone-on-noise and the noise-on-tone condition. Peripheral filters with a steeper high-frequency branch would have to be assumed within the model to account for masking patterns obtained at high masker levels. Moreover, effects produced by the nonlinear signal processing in the auditory periphery influence the masking patterns obtained at high masker levels. The model can not account for these effects because it contains only a linear Gammatone-filter as the initial filtering stage. Experiments on notched-noise masking obtained at a medium masker level are also accounted for by the model. Therefore, the modeling approach of a linear filtering combined with subsequent processing of temporal cues accounts for many aspects of spectral masking with normal-hearing listeners at low and medium stimulus levels.

The general conclusion of this study is to modify the processing model by Dau et al. (1997a+b) for simulating normal hearing by introducing a nonlinear filterbank (or a nonlinear transmission-line model) which realizes a fast-acting compression as the initial (cochlear) processing stage. In a subsequent stage, prior to the nonlinear adaptation stage, an instantaneous expansion (either depending on level or not) has to be incorporated which compensates for effects on stimulus level, produced by the initial compression for normal-hearing listeners and therefore „linearizes" the auditory system. For simulating sensorineural impaired hearing, the compressive properties of the initial processing stage are reduced. If necessary, the variance of the „internal noise" and the width of the peripheral-filter should be increased for simulating the perception abilities of individual hearing impaired subjects.

Appendix A:
Auditory processing model

In Dau et al. (1997a) a model was proposed to describe modulation-detection and modulation-masking data in normal-hearing listeners. The model consists of several preprocessing stages and an optimal detector as decision device (see Fig. A.1). The first processing stage is the linear Gammatone-filterbank model of Patterson et al. (1987), originally designed to account for spectral-masking data obtained with the notched-noise paradigm. The filter shapes are symmetrical on a linear frequency scale and the bandwidths are independent of stimulus level. At the output of each filter, the stimulus is halfwave rectified and lowpass filtered at 1 kHz, which approximates the envelope of the stimulus at high frequencies. The subsequent stage performs a dynamic adaptation. This stage consists of five subsequent nonlinear adaptation loops (Püschel, 1988) with time constants ranging from 5 to 500 ms. Within this stage, stationary signals are compressed with an approximately logarithmic characteristic whereas fast fluctuations (i.e. faster than the time constants of the adaptation loops) are transmitted nearly unchanged. After the adaptation stage, the signal is further processed by a linear modulation filterbank with two domains with different scaling of the filter bandwidth. In the range from 0-10 Hz, a constant bandwidth of 5 Hz is assumed. The lowest modulation filter represents a second-order low-pass filter with a cut-off frequency of 2.5 Hz. From 10 Hz up to the highest modulation filter, a logarithmic scaling with a constant Q-value of 2 is assumed (Dau et al. 1997a). In a more recent implementation of the model, the center frequency of the highest modulation filter was set to a quarter of the center frequency of the peripheral filter (Verhey, 1998). This slight modification of the model was motivated by physiological findings by

Appendix A: Auditory processing model

basilar - membrane filtering

halfwave rectification

lowpass filtering

absolute threshold

adaptation

modulation filterbank

decision device — optimal detector

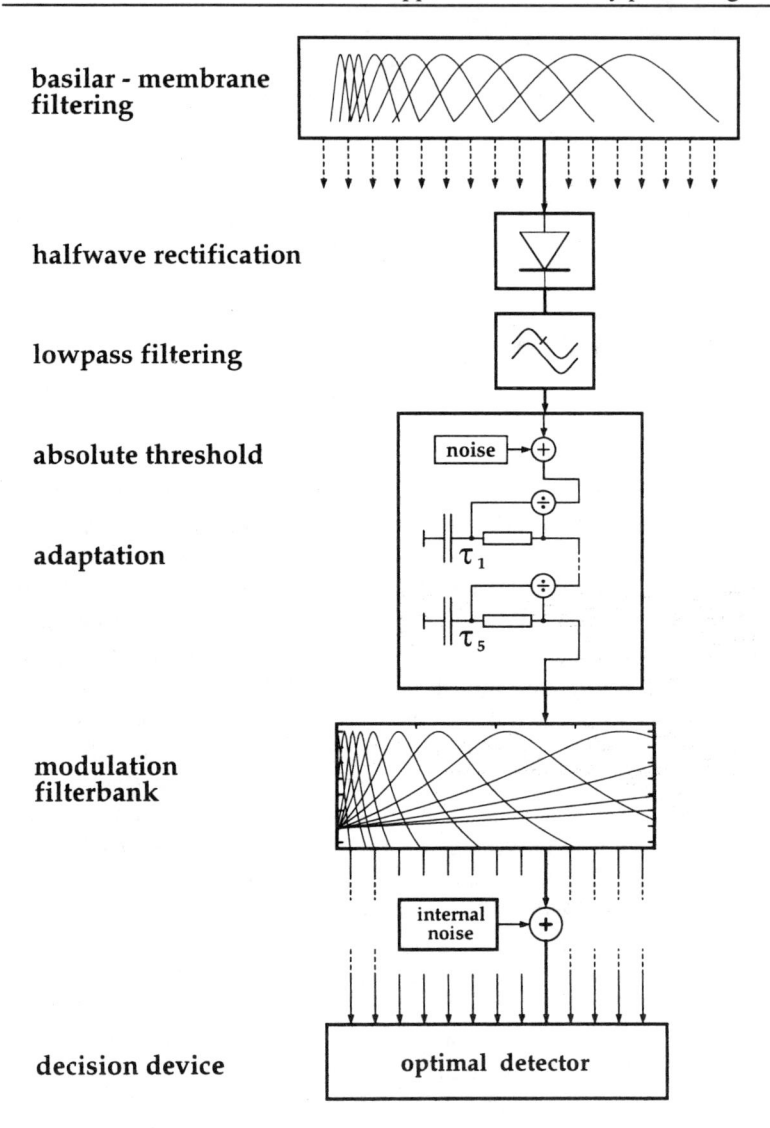

Fig A.1: Model structure.

Langner and Schreiner (1988). All other parameters of the model are identical to those suggested by Dau et al. (1996a+b, 1997a+b). To limit the resolution within the model, an internal noise with constant variance is added to the output of each modulation filter. The subsequent decision stage is realized as an optimal detector, which performs a cross correlation between a „template" and an actual stimulus representation (for details see Dau et al. 1996a+b, 1997a+b). No specific alterations for modeling impaired hearing are introduced within the model.

Appendix B:
Inter- and intraindividual differences

Figure B.1 shows the interindividual (solid line) and intraindividual (dashed lines) standard deviations for the $\Delta f = 200$-Hz condition (upper panels) and for the $\Delta f = 16$-Hz condition (lower panels). The ordinate represents the standard deviation of the modulation depth at threshold, in dB. The abscissa represents modulation frequency. The left panels show the data for the normal-hearing listeners and the right panels for the sensorineural hearing-impaired listeners, respectively. In the $\Delta f = 200$-Hz condition the intraindividual and interindividual data both for the normal-hearing and for the hearing-impaired listeners lie within the range of 3.5 dB and do not depend on modulation frequency. This also holds for the interindividual data in the $\Delta f = 16$-Hz condition. However, the interindividual standard deviation increases markedly at high modulation rates for both groups of subjects. This indicates that the detection ability for high modulation frequencies differs considerably among subjects independent from any hearing loss.

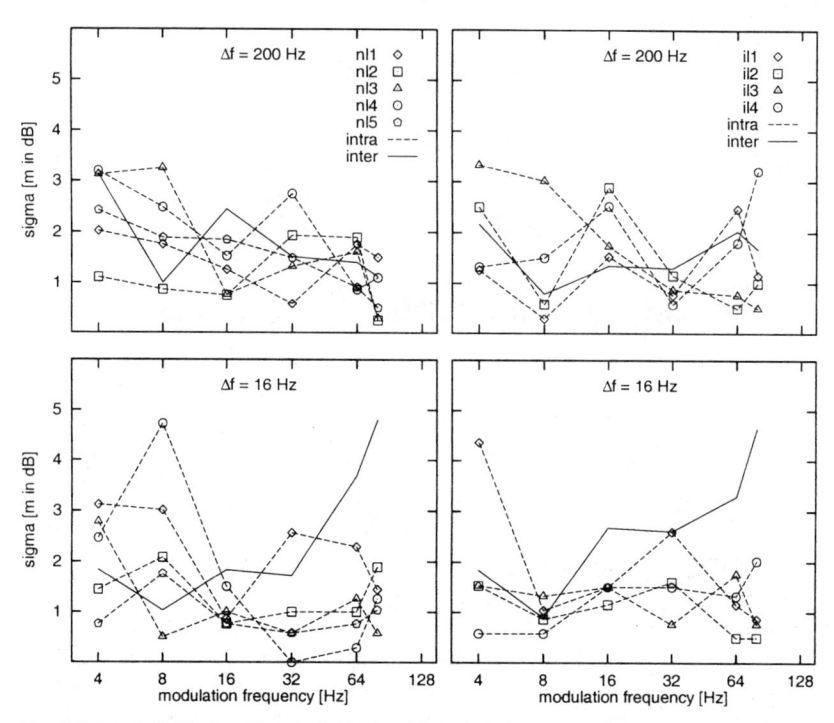

Fig. B.1: Intraindividual and interindividual standard deviation of the modulation detection thresholds for the normal-hearing listeners (left panels) and the sensorineural hearing-impaired listeners (right panels) for the $\Delta f = 200$-Hz condition (upper panels) and for the $\Delta f = 16$-Hz condition (lower panels). The dashed lines indicate the intraindividual and the solid line the interindividual standard deviation.

Appendix C:
Effects of template level

The model version proposed by Dau et al. (1996a) simply derives the template for a threshold estimation at a clearly suprathreshold signal level. It was also pointed out that a more realistic implementation should recompute the template with a decreasing signal level. Such an implementation allows for changes in the detection-cue with decreasing level which is presumably used by a subject. One possible implementation of such an „adaptive" template was used in this study in an iteration procedure. Generally, the suprathreshold level for deriving the template was restricted to a range between 30 and 100 dB. Within this range, a suprathreshold level of the template was chosen well above the expected threshold value, e.g. the measured threshold value, which was adopted from the experiment. Based on this template, a threshold estimate was computed as the first iteration step. For the second iteration step, the suprathreshold level of the template was chosen to be 10 dB above the predicted threshold obtained in the first iteration step. The predicted threshold obtained in the second iteration step was generally lower than the predicted threshold obtained in the first iteration step. Consequently, this iteration was continued until the predicted threshold remained constant, i.e., 10 dB below the template-level. Alternatively, the iteration was stopped if the minimal level for deriving the template was reached.

The adaptation of the template had to be used in the simulation of masking patterns because the predicted threshold depends on the suprathreshold level for deriving the template. This dependency was strongest for signal-masker frequency separations where modulations play an important role within the model. Figure C.1 illustrates this dependency both for tone-masker (solid lines) and noise-masker conditions (dashed lines) for a signal (center) frequency of 900 Hz (masker levels: 85 and 65 dB). The shift of the predicted threshold value is shown as a function of the suprathreshold level for deriving the template relative to the lowest predicted threshold.

The lowest threshold estimates are achieved if the level for deriving the template is 5 to 10 dB above the lowest predicted threshold for both, noise-masker and tone-masker conditions. The range in which the threshold estimate is only little (≤ 2 dB) influenced by the template-level is much broader for the tone-masker (-5 to 20 dB) than for the noise-masker (5 to 15 dB). Outside of these ranges thresholds increase at a rate of approximately ½ dB per change of 1 dB in template level. The cluster of this optimal range in difference between template level and predicted threshold for both maskers amounts to 10 dB. This value was therefore chosen as the target value of the iterative template adjustment procedure. Generally, the iterative procedure converges within 2 to 4 iteration steps to this desired difference.

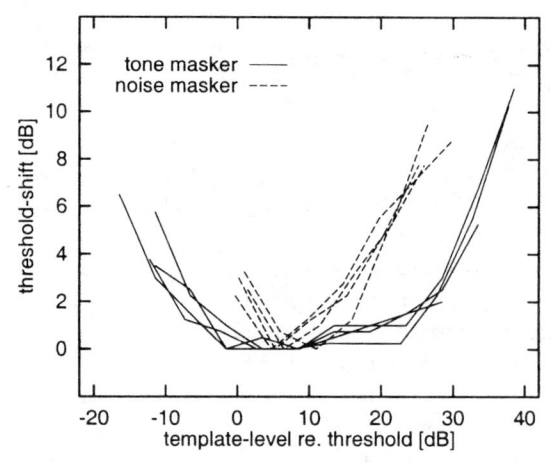

Fig. C.1: The influence of the level for deriving the template (template-level) on the predicted threshold is shown for the tone-masker (solid lines) and noise-masker (dashed lines) condition for the signal center-frequency of 900 Hz. The ordinate indicates the increase in predicted threshold in dB while the abscissa indicates the template-level in dB relative to the lowest predicted threshold estimate. The curves represent model predictions for all signal-masker combinations (**tt, nt, tn, nn**) at masker levels of 65 and 85 dB (not indicated in the figure).

References

Allen, J. B. and Jeng, P. S. (**1990**): „Loudness-growth in 1/2-octave bands (LGOB); a procedure for the assessment of loudness," J. Acoust. Soc. Am. **88**, 745-753.

Bacon, S. P. and Viemeister, N. F. (**1985**): „Temporal modulation transfer functions in normal-hearing and hearing-impaired listeners," Audiology **24**, 117-134.

Boreg, E., Canlon, B. and Engström, B. (**1995**): „Noise induced hearing loss," Scand. Audiol. **24** Suppl. 40.

Brand, T. (**1999**): *Analysis and Optimisation of Psychoacoustical Procedures in Audiology*, PhD thesis, Universität Oldenburg.

Buus, S. (**1985**). "Release from masking caused by envelope fluctuations," J. Acoust. Soc. Am. **78**, 1958-1965.

Carlyon, R. P. and Datta, A. J. (**1997**): „Excitation produced by Schroeder-phase complexes: Evidence for fast-acting compression in the auditory system," J. Acoust. Soc. Am. **101**, 3636-3647.

Carney, L. H. (**1993**) „A model for the responses of low-frequency auditory-nerve fibers in cat," J. Acoust. Soc. Am. **93**, 401-417.

Colburn, S. Zurek, P. and Durlach, N. (**1987**): „Binaural directional hearing - impairments and aids," in *Directional Hearing*, edited by W. Yost and G. Gourevitch (Springer Verlag, New York), 261-278.

Dau, T, Verhey, J. L. and Kohlrausch, A. (**1999**): „Intrinsic envelope fluctuations and modulation detection thresholds for narrowband noise carriers," submitted to J. Acoust. Soc. Am.

Dau, T., Kollmeier, B. and Kohlrausch, A. (**1997a**). „Modeling auditory processing of amplitude modulation: I. Detection and masking with narrowband carriers," J. Acoust. Soc. AM. **102**, 2892-2905.

Dau, T., Kollmeier, B. and Kohlrausch, A. (**1997b**). „Modeling auditory processing of amplitude modulation: II. Spectral and temporal integration," J. Acoust. Soc. AM. **102**, 2906-2919.

Dau, T., Püschel, D. and Kohlrausch, A. (**1996a**). „A quantitative model of the "effective" signal processing in the auditory system. I. Model structure," J. Acoust. Soc. AM. **99**, 3615-3622.

Dau, T., Püschel, D. and Kohlrausch, A. (**1996b**). „A quantitative model of the "effective" signal processing in the auditory system. II. Simulations and measurements," J. Acoust. Soc. AM. **99**, 3623-3631.

Derleth, R. P., Dau, T. and Kollmeier, B. (**1999**): „On the role of envelope modulation processing for spectral masking effects," submitted to J. Acoust. Soc. Am.

Dreschler, W. A. and Plomp, R. (**1980**): „Relations between psychophysical data and speech perception for hearing-impaired subjects. I," J. Acoust. Soc. Am. **68**, 1608-1616.

Dreschler, W. A. and Plomp, R. (**1985**): „Relations between psychophysical data and speech perception for hearing-impaired subjects. II," J. Acoust. Soc. Am. **78**, 1216-1270.

Drullman, R. Festen, J. M. and Plomp, R. (**1994**): „Effects of temporal smearing on speech reception," J. Acoust. Soc. Am. **95**, 1053-1364.

Eddins, D. (**1993**): „Amplitude modulation detection of narrow-band noise: Effects of absolute bandwidth and frequency region," J. Acoust. Soc. Am. **93**, 470-479.

Egan, J. P. and Hake, H. W. (**1950**): "On the masking pattern of a simple auditory stimulus," J. Acoust. Soc. Am. **22**, 622-630.

Ehmer, R. H. (**1959**): „Masking patterns of tones," J. Acoust. Soc. Am. **31**, 1115-1120.

Festen, J. M. and Plomp, R. (**1983**): „Relation between auditory functions in impaired hearing," J. Acoust. Soc. Am. **73**, 652-662.

Fletcher, H. (**1940**): „Auditory patterns," Rev. Mod. Phys. **12**, 47-65.

Fletcher, H. and Munson, W. A. (**1937**): "Relation between loudness and masking," J. Acoust. Soc. Am. **9**, 1-10.

Florentine, M. and Buus, S. (**1984**): „Temporal Gap detection in sensorineural and simulated hearing impairment," J. Speech Hear. Res. **27**, 449-455.

Florentine, M., Buus, S., Scharf, B. and Zwicker, E. (**1980**): „Frequency Selectivity in normally-hearing and hearing-impaired observers," J. Speech Hear. Res. **23**, 646-669.

Forrest, T. G. and Green, D. M. (**1987**): „Detection of partially filled gaps in noise and the temporal modulation transfer function," J. Acoust. Soc. Am. **82**, 1933-1943.

Fowler, E. P. (**1936**): „A method for the early detection of otosclerosis," Arch. Otolaryngol. **24**, 731-741.

Glasberg, B. R. and Moore, B. C. J. (**1989**): „Psychoacoustic abilities of subjects with unilateral and bilateral cochlear hearing impairments and their relationship to the ability to understand speech," Scand. Audiology Suppl. **32**, 1-25.

Glasberg, B. R., Moore, B. C. J. and Bacon, S. P. (**1987**): „Gap detection and masking in hearing-impaired and normal-hearing subjects," J. Acoust. Soc. Am., **81**, 1546-1556.

Greenwood, D. D. (**1971**): "Aural combination tones and auditory masking," J. Acoust. Soc. Am. **50**, 502-543.

Hall, J. L. (**1997**): „Asymmetry of masking revisited: Generalization of masker and probe bandwidth," J. Acoust. Soc. Am. **101**, 1023-1033.

Hansen, M. and Kollmeier, B. (**1997**): „Using a quantitative psychoacoustical signal representation objective speech quality measurements," in *Proc. ICASSP'97, München, Vol. II*, pp.1387.

Hellbrück, J. and Moser, L. (**1985**): „Hörgeräte Audiometrie: Ein computerunterstütztes psychologisches Verfahren zur Hörgeräteanpassung," Psychol. Beiträge **27**, 494-509.

Heller, O. (**1985**): „ Hörfeldaudiometrie mit dem Verfahren der Kategorienunterteilung (KU)," Psychol. Beiträge **27**, 478-493.

Hohmann, V. (**1993**): *Dynamikkompression für Hörgeräte — Psychoakustische Grundlagen und Algorithmen*, Fortschr.-Ber. VDI Reihe **17** Nr. 93. Düsseldorf: VDI-Verlag 1993.

Hohmann, V. and Kollmeier, B. (**1995**): „Weiterentwicklung und klinischer Einsatz der Hörfeldskalierung," Audiol. Akustik, **34** (2), 48-64.

Holube, I. and Kollmeier, B. (**1996**): „Speech intelligibility prediction in hearing-impaired listeners based on a psychoacoustically motivated perception model," J. Acoust. Soc. Am. **100**, 1703-1716.

Hou, Z. and Pavlovic, C. V. (**1994**): „Effects of temporal smearing on temporal resolution, frequency selectivity and speech intelligibility," J. Acoust. Soc. Am. **96**, 1325-1340.

Kissling, J., Steffens, T. and Wagner, I. (**1993**): „Untersuchungen zur praktischen Anwendbarkeit der Lautheitsskalierung," Audiol. Akustik **4/93**, 100-115.

Kohlrausch, A. (**1993**): Comment on „Temporal modulation transfer functions in patients with cochlear implants" [J. Acoust. Soc. Am. **91**, 2156-2164 (1992)], J. Acoust. Soc. Am. **93**, 1649-1650.

Kollmeier, B. and Hohmann, V. (**1995**): „Loudness estimation and compensation employing a categorical scale," in *Advances in Hearing Research*, edited by G. A. Manley, G. M. Klump, C. Köppl, H. Fastl and H. Oeckinghaus (World Scientific, Singapore).

Kollmeier, B., Derleth, R.-P. and Dau, T. (**1997**): „Modeling the "effective" auditory signal processing for hearing-impaired listeners," in *Psychophysical and Physiological Advances in Hearing*, edited by Palmer, A. R., Rees, A., Summerfield, A. Q. and Meddis, R. (Whurr Publisher, London).

Langner, G. and Schreiner, C. (**1988**): „Periodicity coding in the inferior colliculus of the cat. I. Neuronal mechanism," J. Neurophysiol. **60**, 1799-1822.

Launer, S. (**1995**): *Loudness Perception in Listeners with Sensorineural Hearing Impairment*. PhD thesis, Universität Oldenburg.

Launer, S. Hohmann, V. and Kollmeier, B. (**1997**): „Modeling loudness growth and loudness summation in hearing-impaired listeners", in *Modeling Sensorineural Hearing loss*, edited by W. A. Jeastedt (Lawrence Erlbaum, Mahwah), 175-185.

Levitt, H. (**1971**): „Transformed up-down procedures in psychoacoustics," J. Acoust. Soc. Am. **49**, 467-477.

Long, G. R. (**1984**): "The microstructure of quiet and masked thresholds," Hearing Research **15**, 73-87.

Mauermann, M, Uppenkamp, S. van Hengel, P. and Kollmeier, B. (**1999**): „Evidence for the distortion product frequency place as source of DPOAE fine structure. I. Fine structure and higher order DPOAE in dependence on the frequency ratio f2-f1," submitted to J. Acoust. Soc. Am.

Meddis, R. (**1999**): „The auditory periphery as a signal processor," in *Psychophysics, Physiology and Models of hearing*, edited by Dau, T., Hohmann, V. and Kollmeier, B. (World Scientific, Singapore), 131-142.

Moore, B. C. J. (**1995**): *Perceptual Consequences of Cochlear Damage*. Oxford University Press, London.

Moore, B. C. J. (**1997**): „Psychoacoustic consequences of compression in the peripheral auditory system," in *7th Oldenburg Symposium on psychological Acoustics*, edited by A. Schick and M. Klatte (BIS Oldenburg), 565-586.

Moore, B. C. J., Shailer, M. J. and Schoonveldt, G. P. (**1992**): „Temporal modulation transfer functions for band-limited noise in subjects with cochlear hearing loss," British Journal of Audiology **26**, 229-237.

Moore, B. C. J. and Glasberg, B. R. (**1987**): „Factors affecting thresholds for sinusoidal signals in narrow-band maskers with fluctuating envelopes," J. Acoust. Soc. Am. **82**, 69-79.

Moore, B. C. J. and Glasberg, B. R. (**1987**): „Formulae describing frequency selectivity as a function of frequency and level, and their use in calculating excitation patterns," Hear. Res., **28**, 209-225.

Moore, B. C. J., Alcántara, J. I. and Dau, T. (**1998**): „Masking patterns for sinusoidal and narrowband noise maskers," J. Acoust. Soc. Am. **104**, 1023-1038.

Moore, B. C. J., Glasberg, B. R. and Birchall, J. P. (**1985**): „Effects of flanking noise bands on the rate of growth of loudness of tones in normal and recruiting ears," J. Acoust. Soc. Am. **77**, 1505-1515.

Moore, B. C. J., Wojtczak, M. and Vickers, D. A. (**1996**): „Effect of loudness recruitment on the perception of amplitude modulation," J. Acoust. Soc. Am. **100**, 481-489.

Mott, J. B. and Feth, L. L. (**1986**): „Effects of the temporal properties of a masker upon simultaneous masking patterns," in *Auditory Frequency Selectivity*, edited by B. C. J. Moore and R. D. Patterson (Plenum, New York).

Nelson, D. A. and Schroder, A. C. (1996): „Release from upward spread of masking in regions of high-frequency hearing loss," J. Acoust. Soc. Am. **100**, 2266-2277.

Oxenham, A. J. and Plack, C. J. (**1997**): „A behavioral measure of basilar membrane nonlinearity in listeners with normal and impaired hearing," J. Acoust. Soc. Am. **101**, 3666-3675.

Pascoe, D. P. (**1978**): „An approach to hearing aid selection," Hear. Inst. **29**, 12-16.

Patterson, R. D. (**1976**): „Auditory filter shapes derived with noise stimuli," J. Acoust. Soc. Am. **59**, 640-654.

Patterson R. D. and Moore, B. C. J. (**1986**): „Auditory filters and excitation patterns as representations of frequency resolution," in *Frequency Selectivity in Hearing*, edited by B. C. J. Moore (Academic, London), 123-177.

Patterson, R. D., J. Nimmo-Smith, I., Holdsworth, J. and Rice, P. (**1987**): „An efficient auditory filterbank based on the gammatone function," paper presented at a meeting of the IOC Speech Group on Auditory Modeling at RSRE, December 14-15.

Patuzzi, R.B. (**1992**): „Effects of noise on auditory nerve fiber response" in *Noise Induced Hearing Loss*, edited by A. Dancer, D. Henderson, R. Salvi and R. Hamernik (Mosby Year Book, St. Louis), 45-59

Pflüger, M. Hoelderich, R. and Riedler, W. (**1997**): „Nonlinear All-Pole and One-Zero Gammatone Filters," Acustica acta acustica **83**, 513-519.

Plack, C. J. and Oxenham, A. J. (**1998**): „Basilar membrane nonlinearity and the growth of forward masking," J. Acoust. Soc. Am. **103**, 1598-1608.

Pluvinage, V. (**1989**): „Clinical measurement of loudness growth," Hear. Inst. **39**, 28-29.

Püschel, D. (**1988**): *Prinzipien der zeitlichen Analyse beim Hören.* PhD thesis, Universität Göttingen.

Recio, A., Rich, N. C., Narayan, S. S. and Ruggero, R. (**1998**): „Basilar membrane responses to clicks at the base of the chinchilla cochlea," J. Acoust. Soc. Am. **103**, 1972-1989.

Rosen, S. and Baker, R. J. (**1994**): „Characterizing auditory filter nonlinearity," Hear. Res., **73**, 231-243.

Ruggero, M. A. (**1992**): „Responses to sound of the basilar membrane of the mammalian cochlea," Curr. Opin. Neurobiol. **2**, 449-456.

Ruggero, M. A. and Rich, N. C. (**1991**): „Furosemide alters organ of Corti mechanics: Evidence for feedback of outer hair cells upon the basilar membrane," J. Neurosci. **11**, 1057-1067.

Steinberg, J. C. and Gardner, M. B. (**1937**): „The dependency of hearing impairment on sound intensity," J. Acoust. Soc. Am. **9**, 11-23.

Tchorz, J. and Kollmeier, B. (**1999**): „A model of auditory perception as front end for automatic speech recognition," J. Acoust. Soc. Am. (accepted).

Tyler, R. S. Summerfield, Q., Wood, E. J. and Fernandes, M. (**1982**): „Psychoacoustic and phonetic temporal processing in normal and hearing–impaired listeners," J. Acoust. Soc. Am. **72**, 740-752.

Van der Heijden, M. and Kohlrausch, A. (**1995**): „The role of envelope fluctuations in spectral masking," J. Acoust. Soc. Am. **97**, 1800-1807.

Van Hengel, P. W. J. (**1996**): *Emissions from cochlear modelling*, PhD thesis, University Groningen.

van Rooij, J. C. G. M. and Plomp, R. (**1990**): „Auditive and Cognitive factors in speech perception by elderly listeners. II: Multivariate analyses," J. Acoust. Soc. Am. **88**, 2611-2624.

Verhey, J. L. (**1998**): *Psychoacoustics of spectro-temporal effects in masking and loudness perception*. PhD thesis, Universität Oldenburg.

Verhey, L. J., Dau, T. and Kollmeier, B. (**1999**): „Within-channel cues in comodulation masking release (CMR): Experiments and model predictions using a modulation-filterbank model," submitted to J. Acoust. Soc. Am.

Viemeister, N. F. (**1997**): „Temporal modulation transfer functions based upon modulation thresholds," J. Acoust. Soc. Am. **66**, 1364-1380.

Wegel, R. L. and Lane, C. E. (**1924**): "The auditory masking of one sound by another and its probable relation to the dynamics of the inner ear," Phys. Rev. **23**, 266-285.

Yates, G. K. (**1990**): „Basilar membrane nonlinearity and its influence on auditory nerve rate-intensity functions," Hear. Res. **50**, 145-162.

Zeng, F. G., Shannon, R. V. and Hellman, W. S. (**1998**): „Physiological Processes Underlying Psychophysical Laws," in *Psychophysical and Physiological Advances in Hearing*, Proceedings of the 11th International Symposium on Hearing, edited by A. R. Palmer, A. Rees, A. Q. Summerfield and R. Meddis (Whurr Publishers, London).

Zeng, F. G. and Turner, C. E. (**1991**): „Binaural loudness matches in unilateral impaired listeners," Q. J. Exp. Psychol. **43A**, 565-583.

Zerbs, C. (**1999**): *Modeling of the effective binaural signal processing in the human auditory system*, PhD thesis, Universität Oldenburg.

Zwicker, E. (**1956**). "Die elementaren Grundlagen zur Bestimmung der Informationskapazität des Gehörs," Acustica **6**, 356-381.

Zwicker, E. and Fastl, H. (**1990**): *Psychoacoustics — Facts and Models* (Springer-Verlag, Berlin).

Danksagung

An dieser Stelle möchte ich allen Menschen herzlich danken, die auf verschiedene Weise zu dieser Arbeit beigetragen haben.
Herrn Prof. Dr. Dr. Birger Kollmeier danke ich für die Ermöglichung dieser Arbeit in einer Arbeitsgruppe mit ausgezeichneten Arbeitsbedingungen und für sein stetes Interesse an der Weiterentwicklung des „Perzeptions-Modells".
Prof. Dr. Volker Mellert danke ich für sein Interesse und für die Übernahme des Korreferates.
Mein besonderer Dank gilt Dr. Torsten Dau für seine motivationsfördernde Art, die freundschaftliche und fachliche Unterstützung sowie für Korrekturen.
Bei allen Mitgliedern der Arbeitsgruppe „Medizinische Physik" und des Graduiertenkollegs „Psychoakustik" möchte ich mich für die angenehme Arbeitsathmosphäre bedanken.
Insbesondere gilt mein Dank den Personen, die an den oft langwierigen und anstrengenden Messungen teilgenommen haben.
Dr. Jesko Lars Verhey danke ich für unermüdliche Diskussionsbereitschaft und das unerschrockene Herangehen an Fragestellungen.
Dr. Volker Hohmann, Dr. Stefan Uppenkamp und Dr. Martin Hansen danke ich für kritische Fragestellungen, Anmerkungen und Diskussionen.
Oliver Wegner danke ich für die stets prompte Hilfe beim Lösen vieler Soft- und Hardwareprobleme.
Auch bei Prof. Dr. Armin Kohlrausch und Dr. Andrew Oxenham möchte ich mich für ihr Interesse und ihre Anregungen zu dieser Arbeit bedanken.
Nicht zuletzt möchte ich den Mitgliedern der Geschäftsstelle Physik und der Geschäftsstelle angewandte Physik, hier insbesondere Regina Condin, für die Hilfe bei verwaltungstechnischen Fragestellungen danken.

Diese Arbeit wurde finanziell unterstützt von der Deutschen Forschung Gemeinschaft (DFG) im Rahmen des Graduiertenkollegs „Psychoakustik".